Low Temperature Chemical Nanofabrication

Low Temperature Chemical Nanofabrication

Recent Progress, Challenges and Emerging Technologies

OMER NUR
Department of Science and Technology (ITN),
Campus Norrköping, Linköping University,
Norrköping, Sweden

MAGNUS WILLANDER
Department of Science and Technology (ITN),
Campus Norrköping, Linköping University,
Norrköping, Sweden

ELSEVIER
William Andrew
Applied Science Publishers

William Andrew is an imprint of Elsevier
The Boulevard, Langford Lane, Kidlington, Oxford, OX5 1GB, United Kingdom
50 Hampshire Street, 5th Floor, Cambridge, MA 02139, United States

British Library Cataloguing-in-Publication Data
A catalogue record for this book is available from the British Library

Library of Congress Cataloging-in-Publication Data
A catalog record for this book is available from the Library of Congress

ISBN: 978-0-12-813345-3

For Information on all William Andrew publications
visit our website at https://www.elsevier.com/books-and-journals

Publisher: Matthew Deans
Acquisition Editor: Simon Holt
Editorial Project Manager: Thomas Van Der Ploeg
Production Project Manager: Selvaraj Raviraj
Cover Designer: Greg Harris

Typeset by MPS Limited, Chennai, India

Contents

CHAPTER 1
Introduction

1.1 When nanotechnology started?

Although the word "nano" is originally derived from the Greek word "nânos" meaning a dwarf, the term *nanotechnology*, which combines the two words nano and technology, was first introduced in 1974 by the Japanese scientist Norio Taniguchi of Tokyo Science University [1]. However, about 15 years before that, specifically in December 1959, the Nobel laureate physicist Richard Feynman described nanotechnology, although not using the term nanotechnology, during a famous popular speech, "There is Plenty of Room at the Bottom," at the annual meeting of the American Physical Society at Caltech, CA, USA. Feynman had looked two decades ahead and had envisioned and described the possible potential of manipulating and controlling matter on a relatively small scale [2]. Specifically Feynman said: "I would like to describe a field in which little has been done but in which an enormous amount can be done in principle. The field is not quite the same as the others in that it will not tell us much of fundamental physics (in the sense of 'what are the strange particles?'); but it is more like solid-state physics in the sense that it might tell us much of great interest about the strange phenomena that occur in complex situations. Furthermore, a point that is most important is that it would have an enormous number of technical applications. What I want to talk about is the problem of manipulating and controlling things on a small scale" [2]. At the time when Feynman presented his popular visionary talk, there were no spectroscopy tools that could enable scientists to easily visualize, control, or manipulate matter at a relatively small scale. It was more than two decades after Feynman's suggestion of the new area of physics, that scientists were able to visualize with high-precision and control/manipulate atoms and clusters. This was due to the development of the scanning tunneling microscope by IBM scientists in the 1980s, which allowed high-precision visualization at scales of the order of the chemical bond length. These experimental analytical tool developments, combined with advances in theory, and the modeling of matter, has allowed the potential of nanotechnology to be seen and its utilization to be within the reach of human capability.

Low Temperature Chemical Nanofabrication
DOI: https://doi.org/10.1016/B978-0-12-813345-3.00001-0

Nevertheless, long before this, particularly in the early 1940s, some research efforts, while working blindly with no imaging/visualization techniques available, performed the characterization of nanomaterials [3]. In 1944 a paper was published where Fuller characterized nanoscale zinc oxide material using stereoscopic electron microscopy combined with the graphical method of crystallographic stereographic projection [3]. In this pioneering (and long and tedious) work, Fuller came to the conclusion that the structure he was studying was in fact named a "fourlings structure" of a few nanometers in size. The fourlings structure of zinc oxide is a structure with a four-leg arrangement in such a way that the plane formed by each pair of legs is perpendicular to the other plane formed by the second pair of the four legs. The reader of this published work will no doubt see the amount of effort required for Fuller to come to his conclusion. In fact such a conclusion can today be reached within a maximum of 15 minutes of investigation using modern imaging spectroscopy methods, like high-resolution scanning electron microscopy, etc.

The mystery of the change of color of a 1600-year-old Roman goblet cup has amazed scientists for centuries and was not resolved until 1990. As seen in Fig. 1.1, the mysterious Lycurgus Cup, which is made of pigmented glass and decorated with metallic rings, changes color as light is shone on it. It also shows a different color depending on the direction from which the person views it. After being thoroughly investigated, scientists came to the conclusion that the pigmentation is actually made of silver and gold nanoparticles (NPs) of sizes down to 50 nm. This mysterious chalice which is displayed at the British Museum, London, was fabricated

Figure 1.1 The mysterious 1600-year-old Roman goblet cup which changes color when light is shone on it.

using a technique similar to what we use today in nanotechnology. The change of color is explained by the vibration of metallic NPs as light falls on them—more details about why nanosized silver and gold have his ambiguous behavior will be explained in Chapter 2, Phenomenon at the nanoscale. The Lycurgus chalice which is indeed an "out-of-place artifact" (OOPArt) has led scientists to consider the Romans as the pioneers of nanotechnology. But does the search end here? The answer is no!.

Other much older findings discovered only recently have suggested that nanomatter might have been processed and fabricated by humans long before the Romans [4]. In 1991 different morphological nanostructures, e.g., spirals, coils, and shafts, dating back to about 300,000 years, were found near the banks of Russia's Kozhim, Narada, and Balbanyu rivers. These findings were discovered at depths between 10 and 40 feet, and their geological stratus indicated that they are between 20,000 and 318,000 years old. There has been an argument that suggested that these tiny objects might have been left from test rockets experiments from a nearby space research station. Nevertheless, reports have proved that these tiny objects are too old to have originated from modern manufacturing at the claimed space station. Further it was also proved that these thousands of years' old tiny nanostructures are of technological origin. In 1996 Dr. E.W. Matvejeva from the Central Scientific Research Department of Geology and Exploitation of Precious Metals in Moscow wrote that, despite being thousands of years old, the components are of a technological origin. Although this discovery has raised a debate which continues today, it suggests that an advanced culture with high technological capabilities might have existed during the ancient Pleistocene era [4]. There are also claims that these tiny nanostructures, which were fabricated using an advanced technology, do in fact originate from extraterrestrial creatures who gave them to humans or they may have been discarded by extraterrestrials. This claim has been stated with no proof apart from the fact that there was no explanation for the existence of such old technologically advanced metal nanostructures.

However we can ask the following question: is the findings of the OOPArts 300,000 years back is the oldest discovered nanomaterials? The answer is definitely no! According to findings using modern spectroscopy imaging and characterization tools, nanomaterials and nanostructures existed long before that, actually since antiquity. Now humans have learned that nature has always been capable of assembling and creating self-assembled nanostructures and is adapted to nanoengineering [5]. This implies that nature has been assembling nanostructures from time immemorial. To mention some examples, let us consider the structure and functionality of some natural old existing nanomaterials. Let us consider the structure and functionality of *allophane* and *smectite* which are nanomaterials of geological and pedological origin [5].

The allophane structure (see Fig. 1.2) [6] is composed of a hydrous aluminosilicate group appearing as an irregular hollow spherical NP. The outer diameter of the allophane is 3.5–5.0 nm, while the wall thickness is of the range 0.7–1.0 nm [8,9]. The allophane as a nanospherical morphology is a pH-dependent clay mineral that has unique characteristics; it carries both negative and positive charge separated by location in the nanosphere [7]. The positive charge is originating from the *aluminol* group located at the pore region of the nanosphere, while the negative charge originates from the *silanol* group located at the inner side of the allophane nanosphere. *Smectite* has a unit particle composed of an *aluminosilicate* layer with sizes ranging from few tenths of a nanometer to a few hundreds of nanometers, width and length, respectively. The thickness of the layer is about 1 nm only [8]. Due to the relatively small size of both allophane and smectite, their specific surface area is relatively quite large (about 700–900 m^2/g); this implies that a teaspoon of allophane will probably have a surface area much larger than a football playing field [8]. The relatively small size and huge surface area of these naturally occurring materials, that is, *allophane* and *smectite*, along with their peculiar charge characteristics enable them to be excellent contamination sorbent materials and they are widely employed in industrial applications to

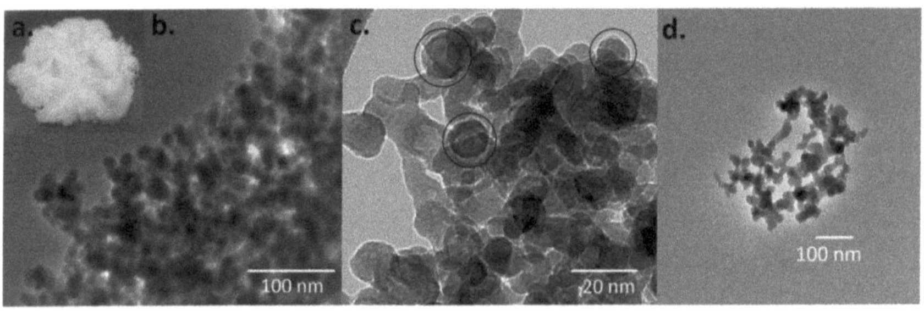

Figure 1.2 Freeze-dried synthetic allophane (A) and the TEM images of the synthetic allophane showing the spherical and hollow morphology of allophane (B and C) as well as allophane nanoaggregates (D). The heat-sensitive allophane was damaged under the electron beam at high magnification, but the consistent spherical shape of allophane and the thickness of the allophane wall (in circles) are evident in photos b and c, respectively [6].

remove pollutants. Beside the use of natural occurring *allophane* NPs in industrial application, they have found their way into medical applications, for example, cytotoxicity of lung cancer [10]. Such old naturally occurring nanostructures have in fact become a source of inspiration for humans to artificially engineer many fascinating new prototypes and useful nanomorphologies of different materials [11]. Another important example of natural occurring nanomaterials is the existence of NPs in natural fresh drinking water [12]. It is now very well established that during evolution living organisms have been shown to be capable of designing biomolecules through self-assembly, building up very smart and complicated organized systems [11]. Hence it is acceptable to say that the existence of nanomaterials is as old as the universe. It is only our relatively recent ability to see, manipulate, and use nanomaterials that has led to the emergence and the early maturity of this fascinating branch of science only recently, that is, during the 21st century. It also worth mentioning that most of the, although not all, naturally occurring nanostructures found today have been evolved at relatively low temperatures ($<100°C$) and as free-standing structures, i.e., no need for a substrate or stand to be utilized to fabricate nanostructures.

1.2 Nanotechnology driving force

It is with no doubt that the discovery and demonstration of the bipolar transistor in 1947 is considered to be the most important discovery of the 20th century [13]. This is due to the huge positive impact of the transistor in our daily lives. The first bipolar transistor was developed using germanium which is a semiconducting material, and was the first three-terminal device that could amplify signals and at the same time be turned on and off allowing control over the passing current. It was Bell Labs scientists John Bardeen, Walter Brattain, and William Shockley who demonstrated the first germanium bipolar transistor and they were awarded the 1956 Nobel Prize in Physics for their discovery. At the same time, and at the General Electronic Laboratory in Paris, France, a working bipolar junction transistor was also demonstrated. The size of this device, i.e., the bipolar transistor, was relatively small compared to the vacuum tubes and heralded the information age. In fact it paved the way for the development of the advanced electronics technology we all benefit from today. Before the transistor, cathode tubes were used and these tubes are relatively large in size and cannot handle high frequencies. The invention of the bipolar transistor was followed by the now dominating metal oxide field effect transistor (MOSFET), ironically demonstrated in 1959 the same year Richard Feynman presented his famous talk about the potential of manipulating and controlling matter at relatively small sizes. Although the field effect transistor idea was anticipated as early as 1926 [14,15], it was, as mentioned above, not until 1959 when Johan Atalla and Dawon Kahng at Bell Labs successfully demonstrated the first working insulator gated field effect transistor [16]. To see how the invention of the transistor has moved us to small-sized smart personal electronics, let us compare the modern small Laptop computers of today to the first developed computer. As mentioned above, before the invention of the transistor, vacuum tubes and other electronics components, e.g., capacitors, were the components used to build electronic systems. The first computer developed was called Electronic Numerical Integrator and Computer (ENIAC) [17]. This relatively large computer occupied a whole large room. The ENIAC construction started in 1946 and by the end of 1955 it contained more than 17,000 vacuum tubes, more than 7000 crystal diodes, about 1500 relays, 70,000 resistors, 10,000 capacitors, and 5,000,000 hand-soldered joining points, with a total weight of more than 25 t, occupying about 72 m^2, and consuming 150 kW of electricity [17]. The ENIAC was then called the "Giant Brain" and was initially intended to calculate artillery firing tables for the United States Army's Ballistic Research Laboratories. At that time the ENIAC provided services that excited scientists and industrial personnel, as ENIAC reduced 20 hours' worth of human hand calculations to about 30 seconds. Nevertheless, today's technology provides a hand calculator that performs the same calculations in a small fraction of a second.

After the establishment of the transistor as an electronic device to control and amplify signals, the development of the electronics industry has been very rapid. The idea to integrate more than one transistor in a single chip, although not technologically achieved yet, has been patented by Jack Kilby of Texas Instruments, and he was awarded the Nobel Prize in 2000 [18]. This was followed by the demonstration of the first planar integrated circuit (IC) the same year. Another breakthrough appeared in 1963 from Sah and Wanlass from Fairchild R&D when they demonstrated the complementary MOSFET (CMOS) which combines p- and n-channel MOSFETs [19,20]. Using the CMOS concept, Sah and Wanlass demonstrated the first logic circuit. At this stage the electronic industry started to mature and in 1965 Moore published a paper describing the development of the number of transistors that will be integrated in a single IC [21]. He projected that the number of transistors in a single IC will double every year for at least a decade. In 1975, a decade later, Moore's law was modified to be that the number of transistors in a single chip will double every 2 years. This is obviously due to the difficulty of miniaturization, i.e., difficulty in shrinking the size of the channel length of the MOSFET [22]. Fig. 1.3 shows the development of the transistor and the following planar IC demonstration. In the first IC only four transistors were integrated together, while today more than 5 million transistors are integrated on a single processor chip. Fig. 1.4 shows the progress of integration since 1971, together with the size order of the active feature of the integrated transistors. As can be seen, long before the year 2000 the feature size of the active device part reached the 100 nm domain. All the amazing development in shrinking the feature size of a single transistor has been achieved by the development of analytical tools to visualize and manipulate matter at relatively small sizes. Since "nanotechnology" is defined as the science, engineering, and technology conducted with materials having at least one dimension which is about 1–100 nm in length, it is then of no doubt that the development of

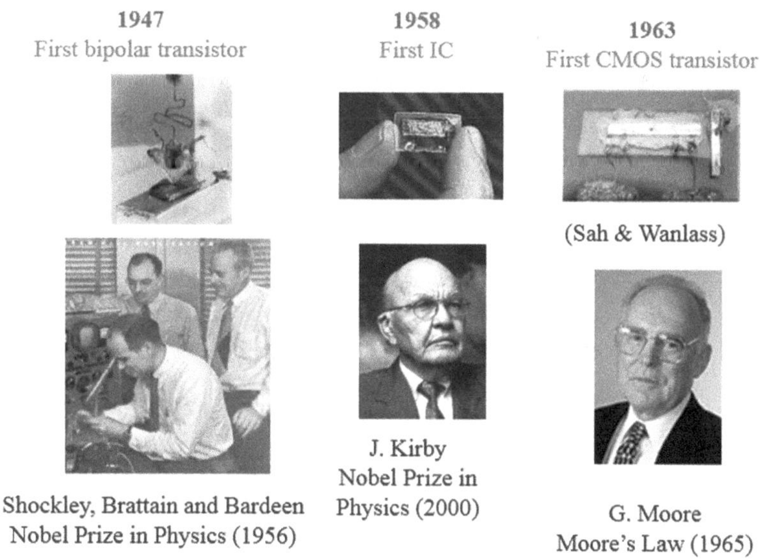

Figure 1.3 The demonstration of the bipolar junction transistor followed by the appearance of the first planar integrated circuit in 1961.

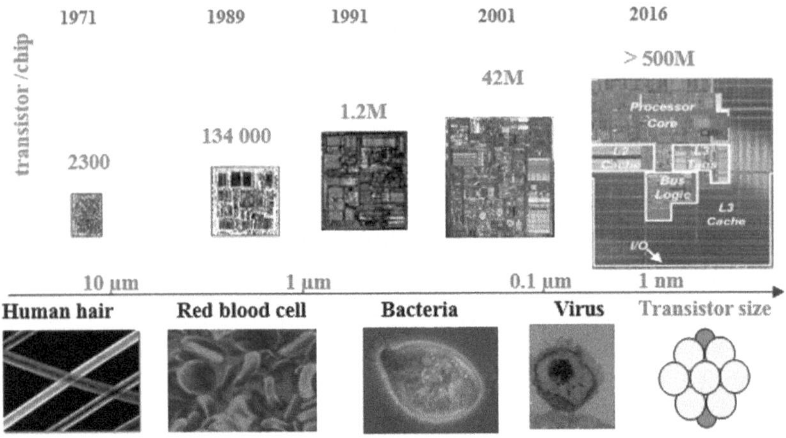

Figure 1.4 The development of integrated circuits with increasing numbers of transistors and the order of size of the active part of the transistor.

microelectronics in pushing the channel length of a MOSFET down below the 100 nm length scale has been the main driving force that led to researchers being able to reach the era of today's nanotechnology. The continuation of the miniaturization of electronic components has reached limits where the planar CMOS-FET channel length has shrunk to just a few nanometers, and consequently devices are no longer operating with conventional performance. This has led to the invention of vertical FET and nanowires-based FET as future alternatives [23]. Although it is the electronics industry that led to equip researchers with analytical tools, today nanotechnology has applications in all fields of science and technology, including engineering, materials science, physics, chemistry, and biology.

1.3 Why nanostructures?

According to the European Commission a nanomaterial is defined as [24]: *A natural, incidental or manufactured material containing particles, in an unbound state or as an aggregate or as an agglomerate and where, for 50% or more of the particles in the number size distribution, one or more external dimensions is in the size range 1 nm–100 nm.* Nevertheless, many published papers are dealing with materials with one dimension in excess of 100 nm that are considered as nanomaterials. The question is: what is special about a "nanomaterial"?

To answer this question, we need to investigate the physical properties of objects when they are scaled isomorphically. With isomorphic scaling we mean that all dimensions are shrunk down in size. If we consider a ball of radius (r), the ratio of the surface area (S) to volume (V) is given by:

$$\frac{S}{V} = \frac{4\pi r^2}{(4/3)\pi r^3} \sim \frac{1}{r} \tag{1.1}$$

Hence the ratio is approximately the reciprocal of the radius. The ratio, which is $\sim 1/r$, increases rapidly as r decreases. Fig. 1.5 shows a Russian nesting doll which is a good illustration of isomorphic scaling and its consequences, as the smallest doll has a relatively much larger ratio of surface area to volume. Although this is a mathematical consideration, the physical meaning will imply that it is no longer the conventional forces that dominate [25]. In principle when an object is isomorphically scaled, i.e., all dimensions are reduced, the change of length, area, and volume ratios will alter the physical observations and even the chemical properties in an unexpected way. Hence if an object shrinks down in size

Figure 1.5 The Russian nesting doll is an illustration of isomorphic scaling. The smallest doll has the largest surface area to volume ratio.

isomorphically and its dimensions reach the boundary effects, i.e., thermal, optical, diffusion, etc., the continuum conventional theories and laws that apply to bulky objects break down and become invalid [25]. So for a shrinking object and when the *S/V* becomes relatively large, the surface forces will dominate over other forces, e.g., gravitational force. In general, when an object is scaled down, the force scaling follows the following rule: forces are scaled for those forces with a lower power of the linear dimension in a way to keep them dominating. A simple example of this is that for an object of mass (*m*), if this object shrinks down and the *S/V* becomes relatively much larger, then the surface tension will dominate over gravity. Since scaling issues are fundamentally important, Chapter 2, Phenomenon at the nanoscale, is devoted to a comprehensive discussion on the phenomenon at the nanometer scale.

1.4 This book

Although this book is devoted basically to highlighting the new emerging low-temperature nanofabrication methods and processes and the associated new emerging applications and technologies, the first chapters are devoted to supporting the main topic of the book as described below.

A brief description of the emergence of nanotechnology will be given, and the accidental early work from the 1940s related to nanofabrication is reviewed. Then other important historical findings related to nanofabrication are stated. This is followed by a discussion on the driving force that led to the era of today's modern nanotechnology rise. Finally, why nanostructures are interesting is highlighted. In Chapter 2, Phenomenon at the nanoscale, the consequences of a reduction in size are reviewed. These include size-related effects, like the modification of the band structure, i.e., bandgap modification, surface effects, and nonlinearity effects. In addition, the modification of different important physical and chemical properties of nanostructures compared to the bulk count partner are reviewed and analyzed. The technological potential for new devices like self-powered technology are introduced (this will be discussed in more detail in Chapter 6: Emerging new applications and future prospects and challenges). In Chapter 3, Conventional nanofabrication methods, the basic different approaches to nanofabrication, i.e., the top-down and bottom-up approaches are discussed with examples. Furthermore, the nanofabrication methods are then discussed as being either physical, i.e., high-temperature, or chemical-based, i.e., low-temperature, approaches.

The rest of the chapter will be devoted to briefly presenting and discussing the different conventional nanofabrication methods. This includes conventional e-beam lithography, and other similar conventional methods as top-down strategies will be discussed. Also other recently developed high-temperature physical approaches following the bottom-up strategy are presented. The advantages, drawbacks, and limitations of these methods are critically discussed. In Chapter 4, New emerging nanofabrication methods, newly developed bottom-up low-temperature methods are introduced as efficient approaches that have potential for developing high-quality nanostructures suitable for many functional devices. Here both top-down and bottom-up strategies performed at both high and low processing temperatures are introduced. In Chapter 5, Low-temperature chemical nanofabrication methods, all the different low-temperature chemical nanofabrication methods are discussed in detail. The hydrothermal low-temperature chemical fabrication of nanostructures will be presented and discussed in detail. This is due to the fact that it is the most important and most common approach, which has been intensively investigated. All the discussion on the hydrothermal low-temperature chemical approach will be accompanied by examples of recent research findings from laboratories worldwide. Finally, the advantages and disadvantages of these methods are discussed critically. In Chapter 6, Emerging new applications and future prospects and challenges, the recently emerging new applications, especially for energy sustainability, are discussed. Here the focus will be on applications related to the low-temperature bottom-up approaches. An example of such an emerging application is the concept of self-powered devices. Examples from recent findings on self-powered bio- and chemical sensors as well as for other mechanical phenomena like piezoelectricity and triboelectricity are also discussed. In addition, recent efforts in developing efficient energy production processes using nanomaterials, e.g., hydrogen production by water splitting, and also the utilization of solar radiation–driven efficient photocatalytic processes are all discussed as promising demonstrations of the potential of the bottom-up low-temperature chemically synthesized nanomaterials. Finally, applications related to human health with energy aspects, like self-powered degradable implanted nanosensors, are introduced. In Chapter 7 the potential of nanomaterials produced using the bottom-up approach, especially following the low-temperature chemical route, in defining new functional devices and introducing electronics in new environments are projected and discussed. The milestones and challenges of the emerging new technologies are critically introduced and analyzed.

References

[1] Taniguchi, N., Proceedings of the International Conference on Production Engineering, Tokyo, Part II. Japan Society of Precision Engineering, 1974.
[2] R.P. Feynman, Eng. Sci. 23 (1960) 22.
[3] M.L. Fuller, J. Appl. Phys. 15 (1944) 164.
[4] L. Vintini, Ancient Nanostructures Found Out of Place and Time, Epoch Times, 2014. <http://www.theepochtimes.com/n3/1058362-ancient-nanostructures-found-out-of-place-and-time/>.
[5] G. Yuan, J. Environ. Sci. Health, A A39 (2004) 2661.
[6] Y-T. Huang, D.J. Lowe, G.J. Churchman, L.A. Schipper, R. Cursons, H. Zhang, T-Y. Chen, A, Cooper, DNA adsorption by nanocrystalline allophane spherules and nanoaggregates, and implications for carbon sequestration in Andisols. Applied Clay Science 120 (2016) 40–50.
[7] A. Ghoneim, N. Matsue, A. Ebi, T. Henmi, J. Appl. Sci. 7 (1) (2007) 103.
[8] S.I. Wada, K. Wada, Caly Miner. 12 (1977) 289.
[9] Wada, K., Proc. Inter. Clay Conf., 1978, Elsev. Sci. Pub. Co. Amesterdam, 1979, p. 537.
[10] Y. Toyota, Y. Matsuura, M. Ito, R. Domura, M. Okamoto, S. Arakawa, et al., Appl. Clay Sci. 135 (2017) 485.
[11] F.M. Fenandes, T. Coradin, C. Aime, Nanomaterials 4 (2014) 792.
[12] C. Latkoczy, R. Kägi, M. Fierz, M. Ritzmann, D. Günther, M. Boller, J. Environ. Monit. 12 (2010) 1422–1429.
[13] W. Shockley, Circuit Element Utilizing Semiconductive Material, United States Patent 2,569,347, 1951.
[14] J.E. Lilienfeld, Method and Apparatus for Controlling Electric Currents, U. S. Patent No. 1,745,175, 1930 (Filed October 8, 1926. Issued January 18, 1930).
[15] J.E. Lilienfeld, Device for Controlling Electric Current, U. S. Patent No. 1,900,018, 1933. (Filed March 28, 1928. Issued March 7, 1933).
[16] D. Kahng, Electric Field Controlled Semiconductor Device, U. S. Patent No. 3,102,230, 1963 (Filed 31 May 31, 1960, issued August 27, 1963).
[17] J.P. Eckert Jr., J.W. Mauchly, Electronic Numerical Integrator and Computer, United States Patent Office, US Patent 3,120,606, filed 26 June 1947, issued 4 February 1964; invalidated 19 October 1973 after court ruling on Honeywell v. Sperry Rand.
[18] J.S. Kilby, U.S. Patent 3,115,581 Miniature Semiconductor Integrated Circuit, filed May 1959, issued December 1963.
[19] F.M. Wanlass, C.T. Sah, Nanowatt logic using field-effect metal-oxide semiconductor triodes, in: International Solid State Circuits Conference Digest of Technical Papers, February 20, 1963, pp. 32–33.
[20] F.M. Wanlass, Low Stand-By Power Complementary Field Effect Circuitry, U. S. Patent 3,356,858, 1967 (Filed June 18, 1963. Issued December 5, 1967).
[21] G.E. Moore, Electronics 38 (8) (1965).
[22] G.E. Moore, Progress in digital integrated electronics, in: International Electron Devices Meeting, IEEE, 1975, pp. 11–13.
[23] R. Xie, C.-C. Yeh, X. Cai, Q. Liu, Series Resistance Reduction in Vertically Stacked Silicon Nanowire Transistors, US Patent: US 14/739543, 2015.
[24] J. Potočnik, Commission recommendation of 18 October 2011 on the definition of nanomaterial, Off. J. Eur. Union 54 (2011). Available from: https://doi.org/10.3000/19770677.L_2011.275.eng. L 275/38.
[25] M.J. Madou, Scaling issues in chemical and biological sensors, in: Proc. IEEE 91, 2003, pp. 830–838.

CHAPTER 2
Phenomenon at the nanoscale

2.1 Why a nanomaterial is different from the bulk?

Materials have been known to humans since the *Stone Age*, when natural materials like stone, wood, skin, etc. were utilized for different purposes. The first nonnatural material was utilized in the *Bronze Age*, dating back to about 3000 BCE. The names *Stone Age* and *Bronze Age* associated with the material name indicate the importance of materials in civilization development. The *Bronze age* was then overtaken by the discovery of iron and steel at around 1200 BCE as being strong materials that can help humans to develop many useful tools. It was more than 2000 years later, at around CE 1850, when low-cost fabrication methods were developed to obtain steel in a cheap way, and then relatively modern revolutionary developments like railways etc. took place in order to serve humans on Earth [1]. Today we live in the *nanomaterials era*. Where revolutionary products for almost all kinds of applications have already appeared in the market or are expected to be available soon. The recent research on nanomaterials, which started about 25 years ago, is at the moment providing fruitful results and is helping to revolutionize a variety of technological and industrial sections, for example, medical tools and drug delivery routines, food control, environmental science, homeland security, and developing new energy sources alternatives.

In general, materials can be classified according to different properties. For example, if we consider electrical conductivity, materials can be classified into metals, insulators, and semiconductors. As mentioned above, and although nonnatural materials have been known to humans since the *Bronze Age*, the understanding of the different behaviors of different materials and why materials are different in response to different external force, that is, mechanical, magnetic, electrical, etc., was only possible relatively recently. It was only during the 1930s that scientists were able to explain the different behavior of different materials. This understanding was possible due to the atomic level explanation by quantum mechanics. In view of the quantum mechanics understanding of atoms, the properties of solids could be understood and the differences were well explained. The properties

Low Temperature Chemical Nanofabrication
DOI: https://doi.org/10.1016/B978-0-12-813345-3.00002-2

of bulk materials, that is, physical, chemical, mechanical, optical, magnetic, etc., are basically due to the band structure theory. The band structure theory defines the range of energies a carrier, for example, electrons, can or cannot have in a solid. According to band structure theory, the energy value of an electron can have a discrete nature, that is, the energy of electrons is quantized. So the energies of electrons in a solid have values that are collected in different groups called bands. At the same time, there are energy values that are not allowed for electron occupancy. These nonallowed energy values define the so-called bandgap or forbidden band. These findings of the band structure theory are based on examining the allowed quantum mechanical wave functions of an electron in a solid with the assumption of an infinite long-range order arrangement of arrays of atoms forming the solid. Due to the overlap of the quantum mechanical wave functions of electrons, the allowed energy values (bands) or the nonallowed energy values (bandgaps) are derived. The band structure theory has been successful in explaining many of the physical properties of bulk solid materials, and has been the base of understanding all solid-state devices. An important assumption of the band structure theory is the fact that the arrangement of atoms in a solid is an *infinite long-range order*! This is not the case when we consider a nanomaterial with rather relatively small dimensions. Hence it is not wrong to expect that the conventional band structure theory, which is successful in explaining the properties of bulky solids, breaks down and fails to explain the behavior of a solid when considering isomorphically scaled down structures, that is, nanomaterials.

Another important concept for understating the difference between bulk materials and nanomaterials and which is important to elaborate, is the concept of density of states (DOS). The DOS is defined as the number of states per interval of energy that can be available for occupation. Hence a high value of the DOS at a specific energy level implies the availability of many states for occupation by carriers, while a zero value of the DOS means that no state is available for occupation at the energy value. Hence the DOS and carriers statistics concept is very essential for studying bulk as well as relatively small material, that is, nanomaterials. Fig. 2.1 displays the density of sates of different structures. As can clearly be seen as the size of the material changes from bulk, to two-dimensional (2D) to one-dimensional (1D) to zero-dimension (OD), the corresponding DOS is different. In turn, different DOS implies different appearance of properties. The reason for this will be elaborated in the next section.

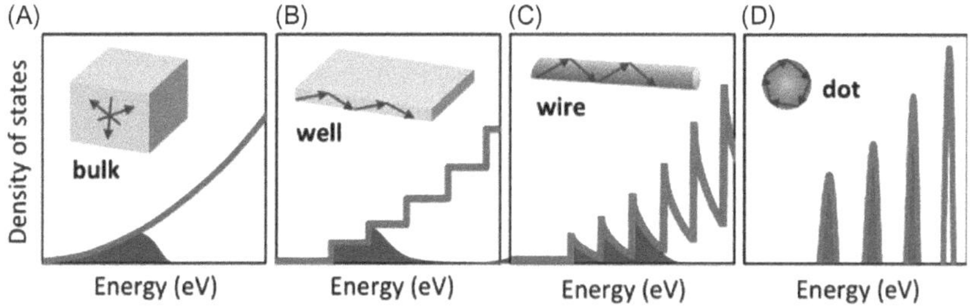

Figure 2.1 (A–D) Schematic diagram of the density of states of different structures.

As mentioned earlier, "when a system is reduced isomorphically in size (i.e., scaled down with all dimensions of the system decreased uniformly, or isomorphic scale reduction), the changes in length, area, and volume ratios alter the relative influence of various physical effects that determine the overall operation—often in unexpected ways. As objects shrink, the ratio of surface area to volume increases, rendering surface forces more important. More generally, as the size of an object decreases, forces scaling with a lower power of the linear dimension dominate over the ones scaling with a higher power (e.g., surface tension gains over gravity, electrostatics over magnetics)" [2]. Hence it is clear that nanomaterials differ from their bulk counterparts basically due to two main reasons, the first is physical (limited number of atoms) and the second is geometrical (due to the significant increase of the surface area to volume ratio compared to the bulk case).

In nature and in most of cases the biological processes occur at relatively small scales, that is, at the nanometer scale, and this provide researchers with models that can be viewed to help construct new artificial processes. Such artificial models and processes can then be utilized to enhance research products and develop new efficient products for medical therapeutics, drug delivery, security, chemical photoprocesses, etc. Hence nanotechnology in fact enables us to utilize the excellent and unique physical, chemical, mechanical, magnetic, and optical properties of small-size materials. It is important to mention that almost all naturally occurring nanoscale materials are produced at rather low temperatures. Hence it is evident that to synthesize nanoscale materials low-temperature processes are feasible processes.

In the following sections we will highlight some of the most important phenomena that take place for nanomaterials that at the same time differ from the bulk counterparts of the same material.

2.2 Electronic band structure of nanomaterials

As mentioned above, the main difference between the properties of a nanomaterial and its bulk counterpart is due to two factors. The first is the relatively large surface area to volume ratio and the second is due to the limited number of atoms and the deviation from the conventional band structure theory main assumption, which states that the material under study has an infinite long-range order of atoms. The effect of these two factors on the electronic properties will briefly be presented and discussed in this section.

Nanomaterials are particles with dimensions lying between the bulk solid state and single atom state, and hence their physical and chemical properties are gradually changing from those of bulk materials to those of single molecules as we shrink the size of the particle. As mentioned earlier, nanomaterials possess a relatively large surface area to volume ratio. This implies that the number of surface atoms may be equal to or in most cases more than the number of atoms at the core of the nanoparticle. As can be seen in Fig. 2.2 the percentage of atoms residing on the surface increases as the size is reduced. Hence surface influence can no longer be ignored. Further, by functionalizing the surface of nanocrystals, even more modification of the properties can be achieved [3]. When the surface of the nanoparticle is free from any adsorbed molecules, the surface atoms will be highly unsaturated and their behavior will be completely different from that of the nanoparticle core atoms, which are fully

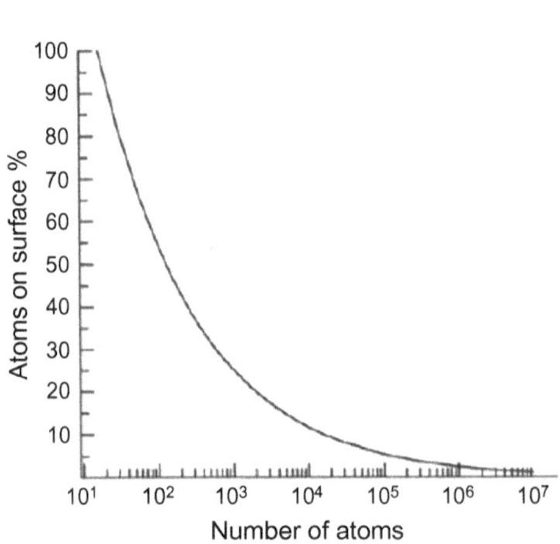

Figure 2.2 The relationship between the number of atoms in a nanocrystal and the percentage of atoms residing on the surface [3].

saturated. If the same case prevails in bulk material, the number of surface atoms will be negligible compared to the number of atoms in the core of the bulk material and hence in this case the surface influence can be neglected. Therefore for the case of nanomaterials and considering the strong influence of surface atoms, different electronic transport and catalysis properties would be expected [4].

As we discuss the electronic properties, the second factor that influences the electronic properties of nanomaterials applies to metals and semiconductors, that is, not for insulating nanomaterials. As the size of a bulk material shrinks down to a particle of nanometer size, the band structure transforms gradually from delocalized bands to become similar to localized molecular orbits band structure. Fig. 2.3 shows a simplified the band structure transition from unfocalized bands to localized molecular orbits when shrinking the dimensions of a bulk material down to the nanometer regime. A typical band structure of a metal is characterized by aquasicontinuous DOS in the valence and conduction band split into discrete energy levels when shrinking the size of a bulky metal down to the nanometer size. The spacing between these discrete energy levels increases with the decreasing metallic particle size [5]. Hence metal particles at nanometer size down to 5 nm or less, that is, metallic particle consisting of 50–100 atoms, transform to behave like semiconductors [6]. Semiconductors characterized by the existence of a bandgap, the situation

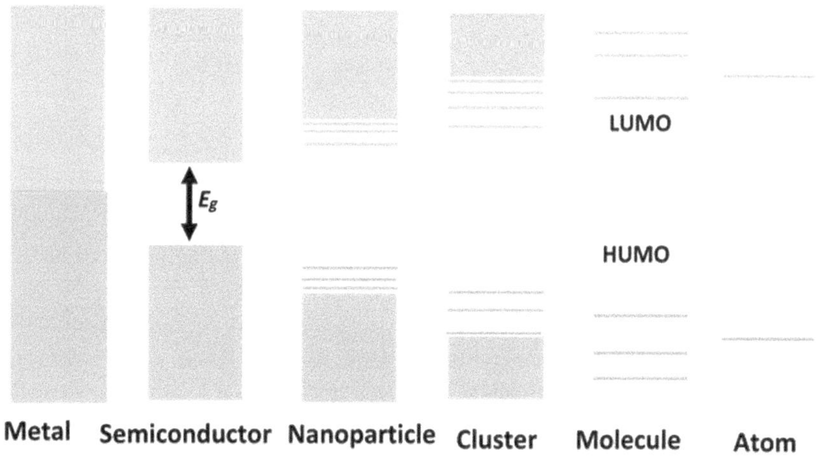

Figure 2.3 Schematic illustration of the increase of the bandgap of a semiconductor as the size of the nanoparticle decreases. As can be seen, nanosized objects possess bandgap properties different from both bulk and atoms.

is rather similar to that of a bulk semiconductor material and as the size shrinks down to the nanometer regime, the bandgap increases as the size decreases and the band structure transforms into discrete energy levels rather than energy bands (see Fig. 2.3). For a typical bulk semiconductor, the band structure can be obtained from the solution of the Schrödinger equation, with the assumption of long-range periodic potential due to the crystal atoms [7]. Further, it is to be noted that the band structure of nanosized materials can change dramatically by slight variations in the morphology of the nanomaterial. An example of this is clearly seen in carbon nanostructures, for example, graphene. Fig. 2.4 displays the band structures of different graphene structures. As can be seen, when changing from a single, to a bi- or trilayer of graphene the band structure is modified [8]. This simple example indicates the potential expected from nanomaterials. However, for a semiconductor nanomaterial and with a reduced size that is less than the de Broglie wavelength, and ignoring the detailed atomic level structure, the energy band levels can be predicted by the

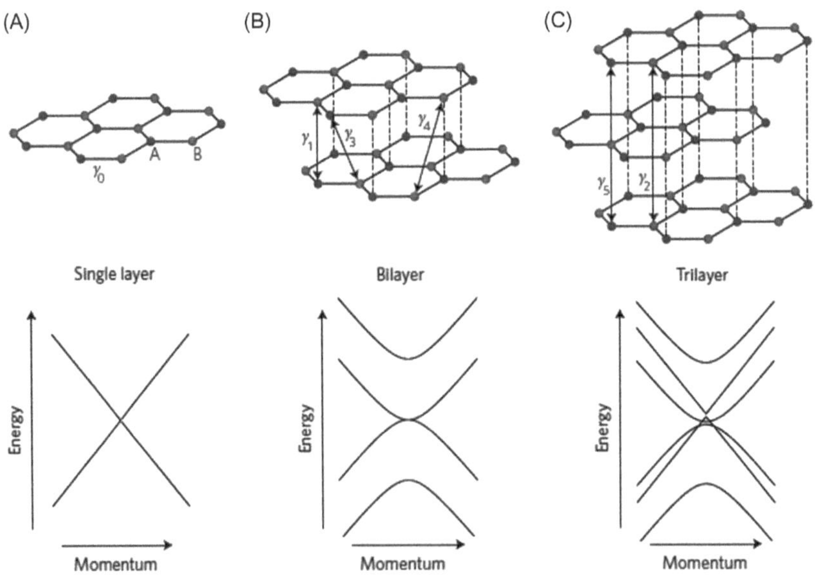

Figure 2.4 (A) Graphene which consists of single layer of carbon, has an electronic structure of conical bands with linear dispersion that touch at a point known as the charge-neutrality, or Dirac, point. (B) On the other hand a bilayer graphene consists of two single graphene layers shifted with respect to each other generating an electronic structure that consists of hyperbolic bands, two of which touch at the charge-neutrality point. (C) Adding a third layer coinciding with the first, the band structure contains both linear and hyperbolic bands [8].

simplified particle in a box problem solved using elementary quantum mechanics [9]. Hence we can generally obtain the energy of a carrier in a 1D box as:

$$E_n = \frac{n^2\hbar^2}{2am_o a} \tag{2.1}$$

where E_n is the energy of the carrier in the nth state ($n = 1$ represent the ground state), ħ is the Planck constant, m_o is the mass of the carrier, and a is the box width, and here it represents the nanomaterial particle dimension, that is, the diameter of a dot in the case of a spherical nanoparticle. So here we have a size quantization effect and the consequences of this effect can be quite useful in developing many new types of electronic devices (see below). As can be seen, moving from bulk semiconductor characterized by a continuous energy bands to nanomaterial we obtain the possible energy levels for a carrier to occupy a set of discrete energy levels. Moreover, as the nanomaterial particle size is reduced further [smaller a in Eq. (2.1)], the ground level increases in value. For the case of bulk crystals, as mentioned above, the energy bands are formed due to the overlap of a very large number of energy levels arising for the large number of atoms or molecules forming the bulky crystal. While in a nanomaterial, which is built of a small number of atoms or molecules, the overlapping of energy orbitals takes place with the available small number of atoms or molecules, and the bandwidths get narrower. This leads to an increase in the energy bandgap of the nanomaterial compared to the bulk counterpart, as schematically shown in Fig. 2.2.

To this end, we have now predicted that a quantization feature is the characteristic of the energy levels of a semiconductor nanomaterial appearing at both sides of the valence and conduction band edges and that the bandgap of this nanomaterial will be larger than its bulk value. Nevertheless, such quantization will have an influence on carrier transport through such relatively small nanostructures. In fact such small structures can be used to control the transport of single or few electrons. Small nanometer-sized structures have been very useful for studying interesting physical phenomenon. If we consider a nanoparticle with a size of a few nanometers, then the number of atoms in this nanoparticle can be about thousands of atoms. For these thousands of atom there exist a similar number of electrons. However, most of these electrons are bound to the nuclei of the atoms. Hence only very few electrons are free and they vary in number from very few to a hundred in a semiconductor quantum dot, that is, in a nanoparticle. These free electors occupy discrete energy levels as described above and possess a wavelength of the order of the de Broglie wavelength. Further, the quantum particles are usually characterized by another physical quantity called charging energy, which is similar to the ionization energy of atoms. The charging energy of quantum semiconductor nanoparticles is defined as the energy required to add or remove an electron from the nanoparticle.

Due to these charging energy characteristics of semiconductor or metallic nanoparticles, these nanoparticles are called "artificial atoms" [9].

The effect associated with the charging effect of quantum dots is called Coulomb blockade [9]. Consider a quantum dot particle surrounded by a source and drain electrodes and connected to a gate, that is, a configuration resembles the metal oxide semiconductor transistor geometry, with no gate-conducting coupling, that is, only capacitive or electrostatic coupling exists. Assuming no coupling to the two electors of the source and drain, then the dot is considered as an isolated island of electrons [10]. Studying the conductance of such configurations reveals a spectacular behavior, that can be used to develop prototype electronic devices and architecture [9].

Assuming that the total number of electrons in the isolated island is N, where N is an integer, the total charge will then be N_e, where e is the electron charge. While tunneling to the source and drain takes place, the total number of electrons N will be adjusted so that the energy of the system is minimum [10]. If the island capacitance is C, then the change of the Coulomb energy will be given by $\Delta E = e^2/C$. This change in the charging energy becomes important when it exceeds the thermal energy $k_B T$. Further, to observe the effect of charge quantization in the presence of the discrete energy levels of nanoscale particles, the barriers are assumed to be sufficiently ambiguous [10]. This means that the electrons are either in the source, drain, or in the island. Further, the tunneling will lead to quantum fluctuation of the number N, which is less than the timescale of measurement [10], where the timescale is approximately given by the electron charge divided by the current. This implies a lower value for the tunneling resistance (R_t). The typical time for charging and discharging the island is given by the resistance R_t multiplied by the island capacitance (C), that is, $\Delta t = R_t C$. By considering the Heisenberg uncertainty relation ($\Delta E \Delta t > h$) and the energy relation shown above we have $R_t > h/e^2$. This implies that the lower limit of the island resistance is given by

$h/e^2 = 25.813$k. Hence [11,12] give the conditions that must be satisfied in order to observe electrical quantization effects in s small-sized object:

$$R_t \gg \frac{h}{e^2} \tag{2.2}$$

$$\frac{e^2}{C} \gg k_B T \tag{2.3}$$

To satisfy the first condition (2.2) the islands should be weakly coupled to the source and drain, while satisfying the second condition requires a reactively small island capacitance and hence it can be satisfied by a relatively small-sized island. It is to be noted that tunneling of a single charge to the island will change the electrostatic energy of the island by a discrete value. On the other hand, if a voltage V_g is applied to a gate of capacitance C_g weakly coupled to the island, it will change the electrostatic energy of the island by a continuous value, then the electrostatic charge of the island will change by a continuous value of $q = C_g V_g$ regardless of the discrete nature of the electronic charge [10]. Now by sweeping the gate voltage, the induced charge will be compensated in a periodic discrete nature through tunneling of discrete charge to the island. The competition between the continuous charge added by the gate and the discrete compensation will lead to the so-called Coulomb blockade phenomenon [12]. Hence when conditions (2.2) and (2.3) are satisfied Coulomb blockade is observed [10]. A variety of new devices that provide exotic electrical characteristics due to the quantization of small size islands have been reported [11]. Among them is a class called single electron transistors (SETs) [11].

The SET is a switching type device, which relies on electron tunneling and can be utilized to amplify a flowing current. In a typical SET there are two tunnel junctions separated from a common metal electrode by a relatively thin (few nm) insulator. Electrons can tunnel through the thin insulator. As tunneling is a discrete phenomenon, the flow of current due to tunneling will be such that its magnitude is a multiple of a single electron charge that is, a multiple of *e*. An example of the Coulomb blockade characteristics observed in an SET is shown in Fig. 2.5 [13]. The set configuration in this example is fabricated using Au for the electrodes and the island, and CrO_x microstrips (see Fig. 2.5 top). The SET was fabricated using two steps of electron-beam lithography and metal

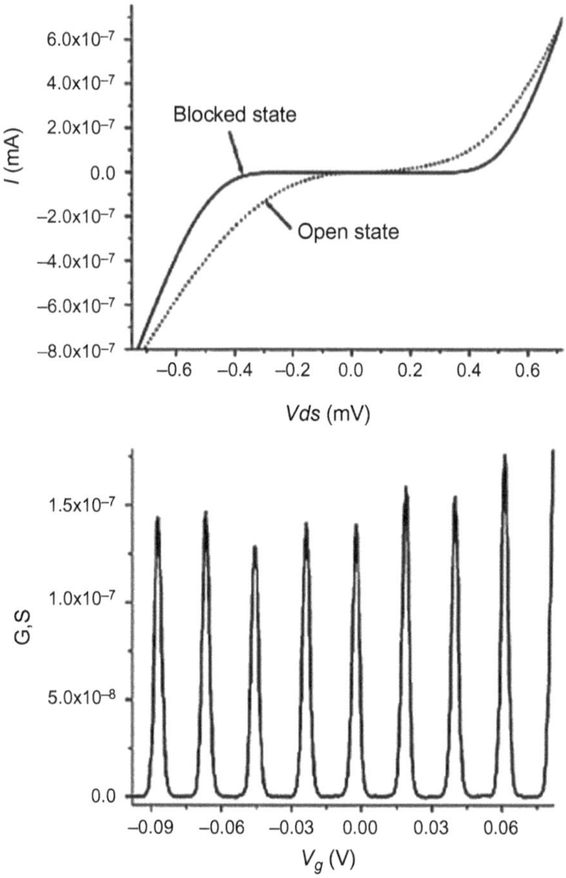

Figure 2.5 (Top) $I-V$ curves of an SET in open state and blocked state. (Bottom) $I-V_g$ modulation curve of the same SET of same SET measured at 300 m K showing deep modulation by the gate. Charging energy is about 0.4 meV. *SET*, Single electron transistor.

deposition [13]. These were chosen to vary the resistance of the SET configuration and obtain matching parameters to satisfy condition (2.2) discussed above. By varying the oxygen content while depositing the Cr, the thinness of the CrO_x could be varied. For the characteristics shown in Fig. 2.5, the charging energy of the SET shown in Fig. 2.5 is about 0.4 meV and about 5 M MΩ resistances at the "open" state (when the Coulomb blockade is lifted) at 300 m K. The $I-V$ curve (Fig. 2.5 middle) and the $I-V_g$ (Fig. 2.5 bottom) show clear SET characteristics with nonlinear open-state behavior [13].

2.3 Optical properties at the nanoscale

Since materials are divided into metals, insulators, and semiconductors, we will discuss the optical properties of each category separately. We start with semiconductors, as they are the most widely used due to their many applications in different types of electronic devices and systems. We then discuss the optical properties of metals at the nanoscale and finally we will say some words about insulating nanomaterials (Figs. 2.6 and 2.7).

As discussed above the decreasing and isomorphic shrinking of the dimensions of semiconducting objects down to the nanometer regime leads to two important consequences. These are (1) an increase of the bandgap as the size is reduced and (2) the appearance of discrete energy levels (with different quantum numbers) appearing at both band edges, that is, Fig. 2.8 shows a clear manifestation of the quantization effect and its impact on the optical properties of relatively small-size nanocrystals of CdSe. The Bohr radius for CdSe bulk crystal is around 4.9 nm, and hence all the shown nanocrystals fall within the strong confinement definition. As can be seen, the confinement begins to affect the excitons wave function, inducing changes in the density of electronic states and in the energy level separation and hence the emitted luminescence is observed to be size dependent [15]. On the other hand, the absorption spectra of relatively small-size nanocrystals show also a profound effect. Fig. 2.9 displays the absorption spectra of different sized CdSe nanocrystals. Note that the largest, being about 20 nm in size and larger than the Bohr radius of CdSe, is still displaying a varying wavelength absorption [16]. Here it is shown that wavelengths of absorption can be tuned by the nanocrystal size. This tailoring of the absorption makes these nanocrystals useful for both biological

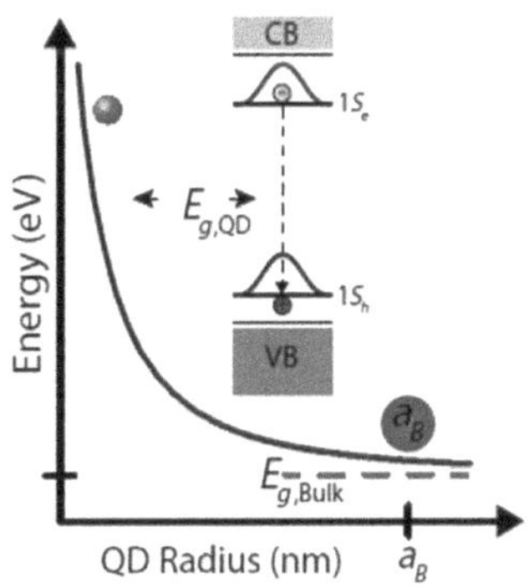

Figure 2.6 Shows the reduction of the bandgap as size decreases [14].

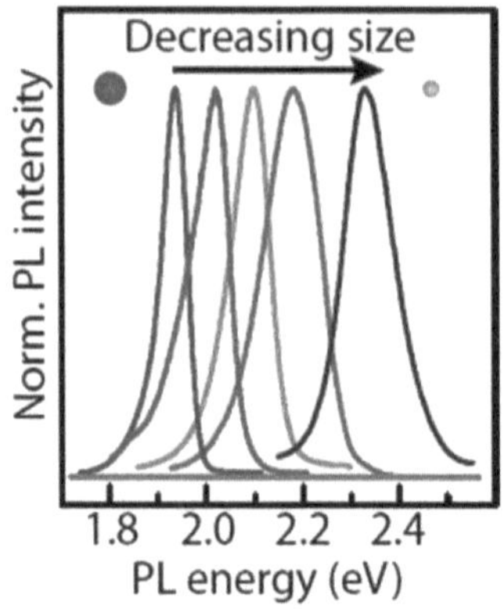

Figure 2.7 Representative PL spectra of QDs ensembles for different nanocrystal sizes. Reducing the quantum dot size shifts the PL emission energy to a higher energy [14].

Figure 2.8 This is an iconic image of different sized (1.7–4.5 nm from left to right) CdSe nanocrystals under ultraviolet illumination. As can clearly be seen, the smallest nanocrystal (1.7 nm) is emitting blue light (longest wavelets compared to the others), while the largest (4.5 nm) is emitting red light (shortest wavelength). This is a clear demonstration of the quantization effect (size-dependent luminescence). Note than the Bohr radius of CdSe is about 4.9 nm.

imaging and optoelectronic applications at the same time; especially since the absorption spectra are rather broad in nature due to the large DOS expected from relatively small-sized nanocrystals like those shown in Fig. 2.9 [16]. In general, semiconductor nanocrystals, being of relatively small size, have found many applications. Due to their light-emitting properties, they have found many applications, including solar energy, optoelectronic devices, cellular imaging, and detectors.

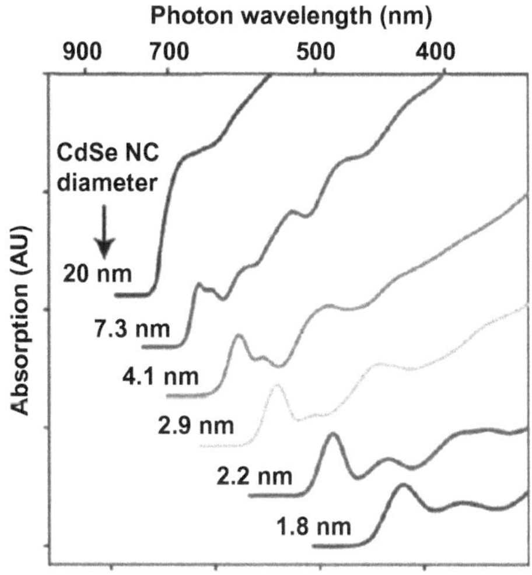

Figure 2.9 Absorption spectra of different sized CdSe nanocrystals [16].

As in semiconducting nanostructures, nanometallic objects also show properties that are quite different from their bulk case. In metals the experimental quantity usually of interest is the conductance (G), which is the current divided by the applied voltage. For a metal strip, the conductance G is given by:

$$G = \sigma \frac{w}{l} \tag{2.4}$$

where w is the strip width and l is its length while σ is the conductivity. If we consider a 3D metal structure, the relation (2.4) is valid if the width w is replaced by the cross-sectional area. However, if l and w are reduced in size down to the nano regime, then relation (2.4) is no longer valid. Novel size dependency will dominate the conductance in that case. Eq. (2.4) is valid as long as we have a diffusive transport, that is, w and l are larger than the mean free path. For smaller dimensions, quantum confinement of carriers in a metallic strip of width w will imply quantization of the energy states available to be discrete. In this case, the energy states are given by:

$$E_n = \frac{n^2\hbar^2}{8wm} \tag{2.5}$$

Hence the number of these width-dependent states which are available for occupation determines the conductance of the nanometallic strip in question. This implies that as the dimensions l and w enter the nanometer regime relation (2.4), a linear dependence on w is no longer valid. It is rather the number of transverse modes that determines the conductance. Hence the quantum mechanics at the nanometer scale force the indirect dependence of G on the width w. As the width of the strip changes, the energy spectrum (2.5) changes and so does the number of occupied modes below the Fermi level, leading to the change of the conductance. Here it is important to note that although they can change continuously and have any value, the number of allowed modes are discrete and have integer values. For the modes the highest occupied mode is given by $E_n = E_F$. Now if we consider the total number of modes to be N, the highest occupied level is

$$N \sim \sqrt{\frac{E_F}{E_n}} \tag{2.6}$$

where E_F is the Fermi energy. Since it is unlikely that the highest occupied level has an eigenvalue that coincides with E_F, then N is accepted to be the integer that corresponds to the highest occupied level below E_F. So by reducing the dimension w, quantum mechanics implies that the conductance changes in a discrete manner following w in a staircase fashion. Physically and at low temperature only, those electrons with characteristic length of the order of the Fermi wavelength (λ_F) can contribute and carry charge. So if $w \gg \lambda_F$, we come to the case of a very large number of conducting transverse modes, and this implies a very large conductance. For metals λ_F is typically around 0.2 nm, which is a relatively small value, and hence the observation of a discrete conductance staircase requires $w \sim \lambda_F$. Fig. 2.10 shows a typical staircase quantum observation of the conductance of a silicon nanowire and a point metallic contact [17,18]. Since the λ_F for most semiconductors is two orders of magnitude larger than metals, discrete conductance behavior was first observed in semiconducting heterojunctions with relatively low electron density. Now considering the effect of reducing the characteristic length l of a conductor, we will see that its effect is even more pronounced than that of the width w. If relation (2.4) holds, that is, Ohm's regime, then as l decreases, we would expect that as l goes to 0, then the resistance should be infinite. Indeed this is not the case. According to the Heisenberg uncertainty principle, the natural characteristic length of an electron (I_e) is defined as the mean free path of the electron. If the length of the conductor is such that $l < I_e$, then electrons can travel through the conductor without losing their initial momentum. This is referred to as a ballistic transport regime. Now following the standard

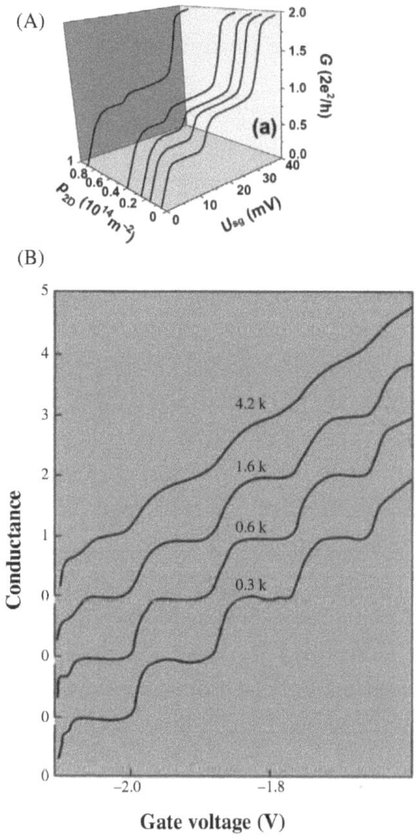

Figure 2.10 Staircase quantum behavior of the conductance of silicon nanowire as a function of the carrier's sheet density controlled through gate voltage [17,18].

definition of conductance as the current passing through divided by the voltage difference, and according to the Heisenberg uncertainty principle, it follows that the electrical conductance is a quantized quantity in nanoscale-sized conductors. Since the current is the charge per unit time and the charge is quantized as an integer of the elementary electron charge e, and considering the extreme quantum limit, the passing charge can be taken as e. Substituting this into the basic definition of the conductance, and considering the potential difference as electrochemical potential difference ΔE divided by the elementary electronic charge, we obtain:

$$G = \frac{e^2}{\Delta E \Delta t} \tag{2.7}$$

Applying the Heisenberg uncertainty principle and considering the spin degeneracy of electrons, we finally obtain:

$$G = \frac{2e^2}{h} \tag{2.8}$$

With conductance in the ideal case of a single electron, this gives a value of $8 \times 10^{-8}\,\Omega^{-1}$. This represents the staircase step height of the conductance per transverse mode. The equivalent resistance (reciprocal of the conductance) has a value of about 12.9 kΩ and it represents the resistance at the contact point where the conductor is connected to the electron reservoir. Hence classically the conductance g is proportional to the width w and considering the quantum mechanical phenomenon, the conductance increases in steps of $2e^2/h$. As the scope of this chapter is to introduce effects at the nanoscale, and for more advanced considerations on the conductance of metallic nanostructures and point contacts the readers are directed to read more in [19,20].

Gold and silver colloidal nanoparticles emit brilliant red and yellow colors, respectively, as seen in the windows of old churches and have been of interest for centuries. The research on the optical properties of metallic nanoparticles dates back as documented to Michael Faraday during the mid-19th century [21]. The first breakthrough that helped in understanding the optical properties was achieved in 1908, when Mie presented an exact solution to Maxwell's equations that described light scattering and absorption, that is, extinction of spherical particles [22]. Although Mie's achievement is still important, it has now been challenged by metallic nanoparticles fabricated by modern methods for applications in medical diagnostics and nanooptics [21]. This is because Mie's theory suits spherical particles, and as it is well documented for many years technology has allowed the fabrication of nonspherical nanoparticles of a variety of shapes using many top-down and bottom-up approaches; some examples of these techniques can be found in [23–26]. Although only semiconductor nanoparticles of sizes down to a few nanometers show strong optical properties with size-dependent modification, metal nanoparticles in access of few 100 nm can show interesting unusual optical properties as will be elaborated below. It is important to mention that it is sometimes complicated to analyze and understand the optical properties of nanoparticles, depending on the surroundings; and this is true whether the particles are spherical or nonspherical [21]. Factors that can lead to complications in understanding the optical properties of nanoparticles can be, for example, the supporting substrate, presence of a solvent layer, or the closeness of nanoparticles causing coupling of their electromagnetic properties, etc. [21]. The emergence of spectroscopy techniques that are quite sensitive to electromagnetic fields close to the surface of the nanoparticle, like surface-enhanced Raman spectroscopy [27], and other nonlinear scattering-based techniques, for example, HyberRayleight [28] and HyberRaman [29], necessitates the need for a general theory, which can accurately describe the electrodynamics of nanoparticles of arbitrary shape and size in the presence of a complex dielectric environment [21]. Such nanoparticles are collectively called anisotropic nanoparticles [30]. Indeed recently different theories that can describe such nanoparticles have been developed

In the last part of this section, we discusses the most important phenomenon of interest for spherical metallic nanoparticles, namely, the surface plasmon resonance.

The free electrons in metals can travel through the metal. If the metal particle size is less than the mean free path, then no scattering is expected to occur. So when a light of wavelength larger than the mean free path falls on the nanoparticle, a standing resonance condition is set. So free electrons in metal nanoparticles will oscillate around the surface. This is why it is called *surface plasmonic resonance*. This is schematically shown in Fig. 2.11. As the light wave passes through, the electron cloud is polarized to one side of the surface and it starts to oscillate. The oscillation frequency depends on the particle size, shape, and surrounding environment. If the dielectric constant of the surrounding material is changed, then the oscillation frequency will change due to the variation of the material's ability to accommodate electron charge density of the electron cloud from the nanoparticle. If the metallic nanoparticle is capped with a different material a shift in the resonance frequency will be observed. This

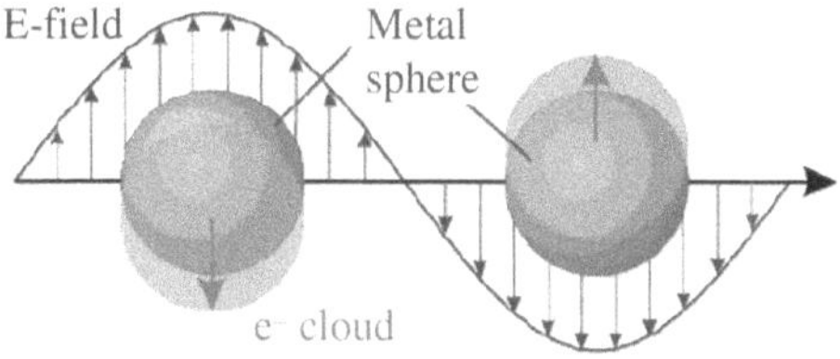

Figure 2.11 Schematic representation of plasmonic oscillations for spherical metallic nanoparticles. The oscillation is in a form of an electron cloud oscillating around the nuclei.

means that by attaching a specific chemical molecule to the surface of a metallic nanoparticle, a specific shift to the frequency of the surface plasmon's absorption maximum is expected. This forms the basis of the use of metallic nanoparticles for sensitive detection of other chemical molecules available near the surface. The recent success in the realization of different nanostructures morphologies, for example, nanorods, nanodisks, nanoprisms etc., has necessitated the development of models and theories to account for their peculiar properties [30]. Due to this, many theories have been developed recently for different nanoparticles in complex environments. Among these are the discrete dipole approximation (DDA), finite difference time domain method, and the modifying long wavelength approximation [31–34]. Among them, the DDA is of particular interest. This is because the DDA can treat particles of arbitrary shape placed in complex environments. In the DDA method, the particle is divided into cubic arrays of N polarizable elements and the response of these element to an applied electromagnetic field is then obtained through self-consistent determination of the response of each cubic element [30]. By employing these theories, the optical response of different anisotropic nanoparticles could be explained by comparing the experimental measurements with theoretical predictions.

Spherical silver and gold nanoparticles show a strong surface plasmon band at around 400 and 520 nm, respectively [35]. Fig. 2.12 shows the extinction spectra of gold nanoparticles with sizes from 4 and up to 99 nm. It is clear that the extinction wavelength is redshifted as the size of the gold nanoparticle size is increased. In Fig. 2.12 the plasmon absorption is clearly visible. In addition, the width follows the expected trend. It is

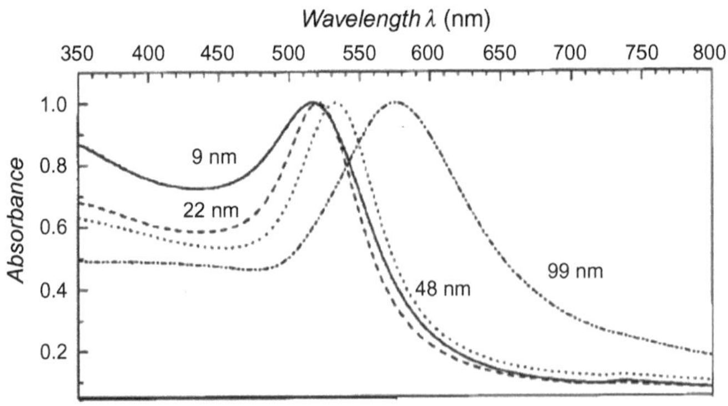

Figure 2.12 Extinction spectra of gold nanoparticles in different sizes.

also seen that intrinsic size effects dominate at the quasistatic limit. Also for relatively larger nanoparticles the plasmon bandwidth increases due to extrinsic effects [35]. Considering spherical nanoparticles as isotropic in nature, Mie theory explains the extinction properties observed from such nanoparticles. Before shifting to examples of optical extinction and surface plasmon's bands observed in some other anisotropic nanoparticles, we take an example of alloying different materials in spherical geometry. Contrary to the weak size-dependent shift of spherical gold nanoparticles' surface plasmon band shown in Fig. 2.12, alloying gold and silver indicates a stronger variation of the plasmon's resonance band [35–37]. Fig. 2.13 shows the absorption spectra of pure gold and silver–gold alloyed spherical nanoparticles. The alloyed nanoparticles of gold–silver were achieved by coreduction of gold and silver ions in sodium citrate [37]. For the composition of the gold fraction $x_{Au} = 0.8$, 0.54, and 0.27, the observed absorption spectra are shown in Fig. 2.13A. As can be seen, as the gold mole fraction is increased in the alloy the maximum of the plasmon's absorption peak is blue-shifted. This is shown in Fig. 2.13B where the plasmon's maximum absorption is plotted versus the gold fraction. In Fig. 2.13B a linear relationship between the blue shift of the plasmon's absorption (increase of the wavelength) and the increase of the gold fraction is evident [38,39]. This rather wide variation in the surface plasmon's absorption band of gold–silver alloyed spherical nanoparticles makes it of interest for many optical applications requiring a specific absorption spectrum due to the possible tuning by varying the fraction of the alloy [35]. The wide variation of possible absorption discussed here is the phenomenon responsible for the different pleasant colors observed in the 1600-year-old Roman goblet cup which has amazed scientists for centuries and which was not resolved until 1990, as discussed in Chapter 1, Introduction. Treating the dielectric content of the alloy through a linear relationship of the two constituents dielectric constant and using this new allot dielectric constant in Mie theory can lead to acceptable estimation of the optical properties of the alloy. This is shown in Fig. 2.13C.

However, anisotropic nanoparticles show different optical extinction and the DDA theory is more appropriate to apply for predicating the optical response of anisotropic nanoparticles of arbitrary shape with different sizes. Below we discuss some examples. Nanorods/nanowires are among the most interesting anisotropic nanoparticles. It is also the most common morphology, as it exists for a broad range of materials. Although

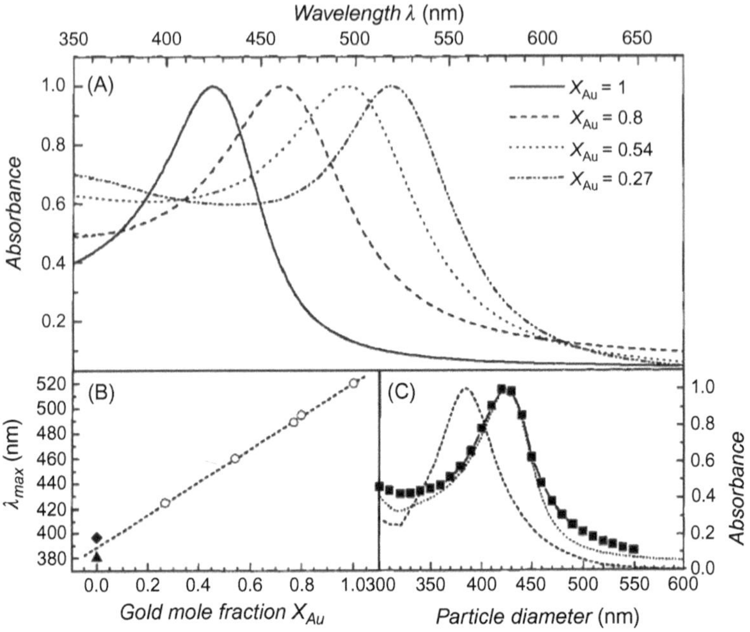

Figure 2.13 Effect of gold–silver alloy formation on the surface plasmon absorption. Part (A) shows the UV–vis absorption spectra of spherical gold–silver alloy nanoparticles of varying composition. The gold mole fraction x_{Au} varies between 1 and 0.27. The plasmon absorption maximum is blue-shifts with decreasing x_{Au}. This fact is further illustrated in part (B) where the plasmon band maximum is plotted as a function of the gold mole fraction x_{Au}. The open circles correspond to experimental data from [36], while the value of the plasmon maximum of pure silver nanoparticles is taken from refs [37,38]. The dashed line is a linear fit to the data points, demonstrating that the dependence of the plasmon band maximum on x_{Au} is best described by a linear relationship. Part (C) shows the experimental UV–vis absorption spectrum of gold–silver alloy nanoparticles (x_{Au}) 0.27 (*dotted line*), and two calculated absorption spectra using the Mie equation and different values for the dielectric function. The dashed line gives the spectrum calculated by using a mole-fraction-weighted dielectric function of gold and silver. Better agreement between theory and experiment is found if the experimental dielectric function obtained from an alloy thin film with a similar composition (x_{Au}) 0.28 is used in the Mie equation (*squares, solid line*).

great success has been achieved in growing nanorods and many other anisotropic morphologies, it is still difficult sometimes to explain all the details of the growth kinetics. Experimental results on the optical distinction of anisotropic nanosized objects, that is, nonspherical nanoparticles like, for example, nanorods, indicated that their properties are drastically changed compared to isotropic nanoparticles, that is, spherical nanoparticles. Fig. 2.14

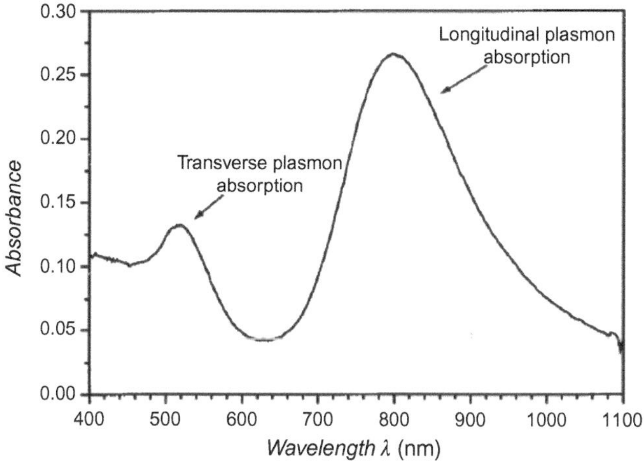

Figure 2.14 The corresponding UV–vis absorption spectrum of colloidal gold nanorods. The absorption band at 520 nm is the transverse mode of the surface plasmon absorption, while the longitudinal mode absorbs around 800 nm for these nanorods with a 4.1 aspect ratio.

displays the corresponding UV–vis absorption spectrum of colloidal gold nanorods.

As clearly seen in this spectrum (Fig. 2.14) and contrary to what is observed from spherical nanoparticles, the corresponding UV–vis absorption spectrum of colloidal gold nanorods indicates two absorption peaks. The observation of two peaks when compared with the effect of size for spherical nanoparticles indicates that the *shape* or *morphology* of nanosized particles can have a more pronounced effect on the observed properties of that nanosized object. In Fig. 2.14 the resonance transverse peak observed at around 520 nm is at the same wavelength as the resonance of spherical Au nanoparticles, while the other peak. called the longitudinal. is red-shifted to a lower wavelength. The position of the longitudinal peak could naturally be dependent on the dimensions of the nanorod, as it is due to resonance along the axis of the nanorods. To investigate the dependence of the longitudinal resonance peak of nanorods observed on the dimensions, that is, the aspect ratio, an extension of Mie's theory can be used [40]. By using the dipole approximation for the extinction coefficient for N nanoparticles in a volume V, and the measured dielectric function of gold, the dependence of the absorption spectra of gold nanorods for different aspect ratio can be calculated [41–43]. For a fixed medium, Fig. 2.15A shows the dependence of the nanorods' absorption

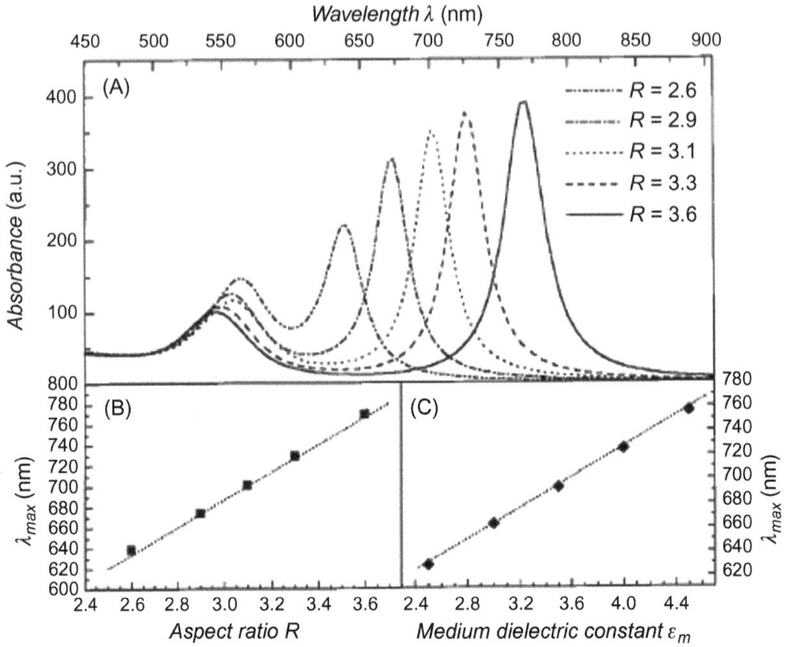

Figure 2.15 Simulation of the surface plasmon absorption for gold nanorods of different aspect ratios. (A) Shows the calculated absorption spectra of gold nanorods with varying aspect ratios of the nanorod. The dependence of the maximum of the longitudinal plasmon absorption as a function of the gold nanorod aspect ratio and the medium dielectric constant for a constant value of the other parameter are given in parts (B) and (C), respectively.

on the aspect ratio *R*. As can be seen in Fig. 2.15A, as the aspect ratio is changed from 2.6 to 3.6 nm, the longitudinal resonance is shifted by about 150 nm. In addition, as can be seen in Fig. 2.15B, the increase in the longitudinal peak position follows a linear trend, as the aspect ratio is increased [35]. For predicting the position of the maximum of the longitudinal resonance, the solid line in Fig. 2.15B can be represented by [43]:

$$\lambda_{\max} = (33.34\ R - 46.31)\varepsilon_m + 427.31 \tag{2.9}$$

where $\lambda_{\max}$ represents the maximum of the longitudinal resonance peak position, *R* is the aspect ratio, and ε_m is the dielectric of the medium. Eq. (2.9) suggested that for a fixed aspect ratio *R*, the maximum of the longitudinal resonance peak position should also have a linear dependence on the dielectric constant of the medium. Fig. 2.16A shows the experimental maximum values of the absorption wavelength of the longitudinal

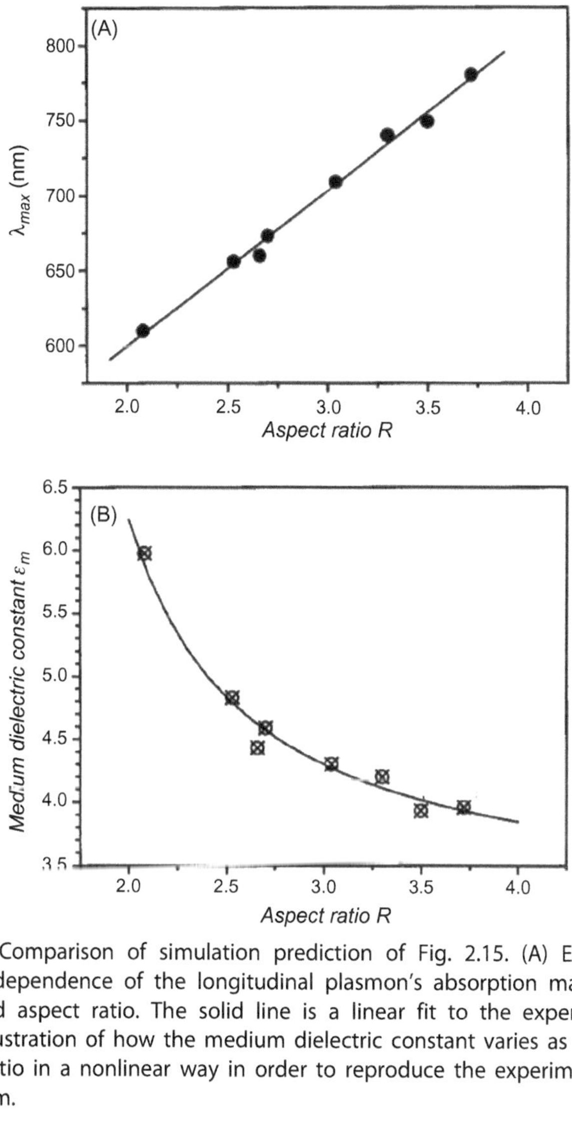

Figure 2.16 Comparison of simulation prediction of Fig. 2.15. (A) Experimentally determined dependence of the longitudinal plasmon's absorption maxima on the gold nanorod aspect ratio. The solid line is a linear fit to the experimental data points. (B) Illustration of how the medium dielectric constant varies as a function of the aspect ratio in a nonlinear way in order to reproduce the experimental absorption maximum.

plasmon resonance of nanorods where a linear relation is clearly observed between the maximum intensity wavelength and the nanorods' aspect ratio [43]. This result was also confirmed by other experiments [44,45]. As Eq. (2.9) also suggested for a fixed aspect ratio the maximum of the plasmon resonance should indicate a linear dependence with the medium dielectric constant, experiments showed a different dependence [35].

2.4 Other nanostructures morphology related effects

Beside the effect of size on the optical properties discussed in the previous section, combining different nanomaterials in composite configurations can further be an approach for possible tuning and altering of the basic optical properties. An example of this is the core–shell nanocrystals (CS NCs) family. In a CS NC, at least two different semiconductor materials are combined in an onion-like structure. By covering a nanoparticle with a thin shell, the core fluorescence, quantum yield, and lifetime can all be tuned for a desired performance [46]. In a CS NCs the shell also provides a protection for the optically active core by passivating the available surface trap states giving rise to further enhancement of the quantum yield.

Passivating the surface traps is particularly important for applications such as biological labeling and light-emitting devices [46]. Hence the shell usually acts as a barrier protecting the core from the environment and this leads to the altering of the charge, functionality, and reactivity of the surface. Further, and as mentioned above it, can change the stability and dispersive ability of the core material [47]. In addition, the electrical, catalytic, optical, or magnetic properties of the core material can be improved by covering it with another material. In general, the synthesis of core–shell structured material leads to a new composite material being obtained with properties between the core and shell materials [47]. A typical example of a CSNP is shown in Fig. 2.17, where a zinc oxide (ZnO) NP is covered with a shell of zinc sulfide. Zinc oxide has a room temperature bandgap of about 3.37 eV, while that of ZnS is about 3.72 eV [47].

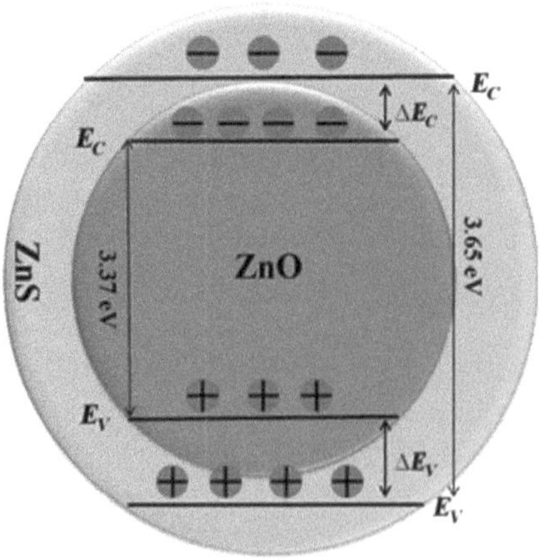

Figure 2.17 Schematic band alignment of ZnO@ZnS CSNPs.

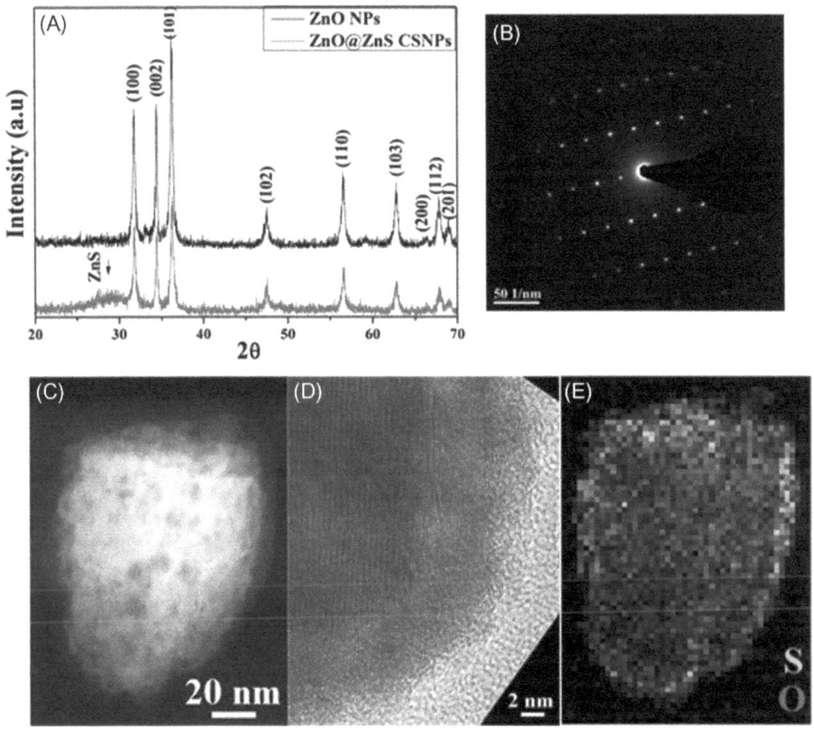

Figure 2.18 (A) XRD spectra of ZnO NPs and ZnO@ZnS CSNPs; (B) corresponding SAED pattern from (A); (C) STEM of ZnO@ZnS CSNPs; (D) HRTEM of ZnO@ZnS CSNPs; and (E) EDX mapping exhibiting S and O distribution.

Fig. 2.18 shows different structural results of ZnO@ZnS CSNPs grown by a chemical low-temperature route (details of the synthesis will be discussed in the following chapters). The ZnO@ZnS CSNP optical bandgap is different from that of both the ZnO and ZnS.

As can be seen in Fig. 2.19, the UV−vis characterization of ZnO NPs and different ZnO@ZnS CSNPs synthesized with different molar ratios have indicated different optical absorption and different optical bandgap [48]. From Fig. 2.19A, the absorption wavelength peak of the pure ZnO NPs is around 360 nm and shifts to a higher wavelength (371 nm) for ZnO@ZnS CSNPs. Fig. 2.19B shows the plots of $(\alpha h\upsilon)^2$ versus $h\upsilon$ for ZnO NPs and ZnO@ZNS CSNPs. The optical bandgap of ZnO NPs is 3.08 eV and it is smaller than the bandgap of the bulk ZnO [48]. The reasons for this is due to the size reduction effect (as discussed above), and in addition to that the presence of structural defects that arise during the preparation of the samples can also contribute to the reduced

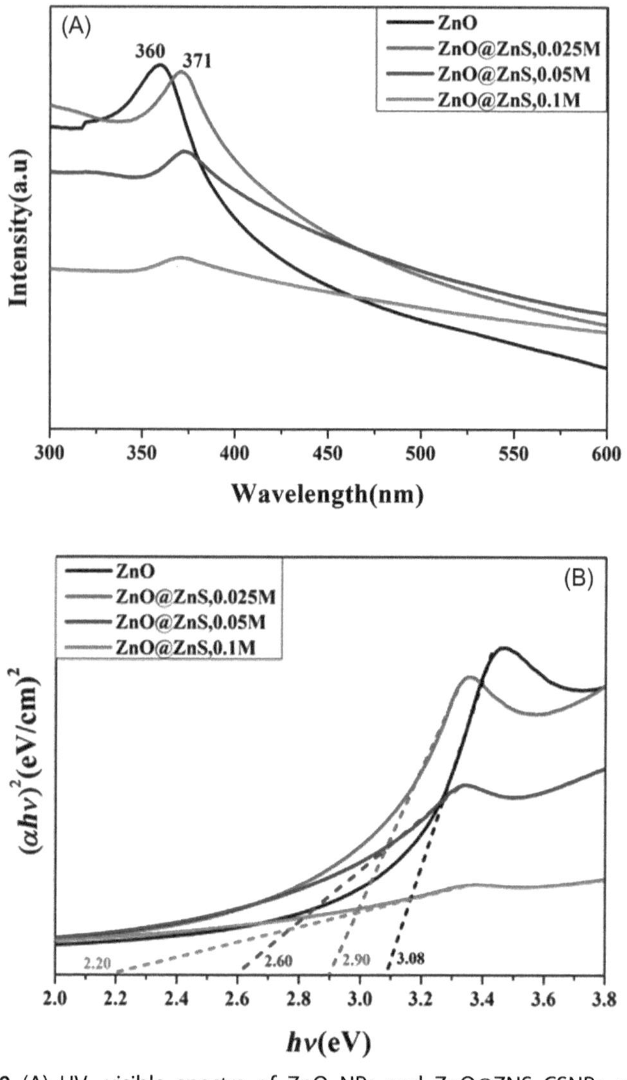

Figure 2.19 (A) UV–visible spectra of ZnO NPs and ZnO@ZNS CSNPs and (B) the plots of $(\alpha h\upsilon)^2$ versus $h\upsilon$ for ZnO NPs and ZnO@ZNS CSNPs.

bandgap [48]. As can be seen, in general, covering the ZnO NPs with ZnS leads to a reduction of the optical bandgap. Here the ZnS shell has been prepared with different values of the concentration of the sulfur source in the nutrient solution. This is indicated in the figure caption as 0.1 M, 0.2 m, and 0.3 M, respectively. As can clearly be seen, as the concentration of the sulfur source is increased the optical bandgap is further reduced. This could be due to the increased coverage of the core by the ZnS [48].

Although the relative band alignment between the core and the shell is important to consider when designing a CSNP, it is not the most important parameter to consider. The most important requirement when synthesizing an optically efficient CSNP is the fact that the shell has to proceed through epitaxial growth over the core material and this means that both the core and the shell has to have the same crystal structure and the relative lattice mismatch between them should be relatively small. Otherwise, a relatively large number of defects, which will act as trapping centers for the photogenerated charged carries and hence reduce the florescence quantum efficiency, will occupy the interface between the shell and the core. Table 2.1 lists selected different semiconductor materials parameters [46]. The crystal structure of the shell should match that of the core to obtain epitaxial growth and avoid defect formation. The second important parameter to consider is the band alignment. This is important because it will dictate what type of band alignment will result from the core–shell configuration. Fig. 2.20 shows the different types of CSNPs

Table 2.1 Some selected semiconductor materials with their bulk parameters, that are of interest for CSNPs design [46].

Material	Structure (300K)	Type	E_{gap} (eV)	Lattice parameter (Å)	Density (kg/m^3)
ZnS	Zinc blende	II–VI	3.61	5.41	4090
ZnSe	Zinc blende	II–VI	2.69	5.668	5266
ZnTe	Zinc blende	II–VI	2.39	6.104	5636
CdS	Wurtzite	II–VI	2.49	4.136/6.714	4820
CdSe	Wurtzite	II–VI	1.74	4.3/7.01	5810
CdTe	Zinc blende	II–VI	1.43	6.482	5870
GaN	Wurtzite	III–V	3.44	3.188/5.185	6095
GaP	Zinc-blende	III–V	2.27	5.45	4138
GaAS	Zinc blende	III–V	1.42	5.653	5318
GaSb	Zinc blende	III–V	0.75	6.096	5614
InN	Wurtzite	III–V	0.8	3.545/5.703	6810
InP	Zinc blende	III–V	1.35	5.869	4787
InAs	Zinc blende	III–V	0.35	6.058	5667
InSb	Zinc blende	III–V	0.23	6.479	5774
PbS	Rocksalt	IV–VI	0.41	5.936	7597
PbSe	Rocksalt	IV–VI	0.28	6.117	8260
PbTe	Rocksalt	IV–VI	0.31	6.462	8219

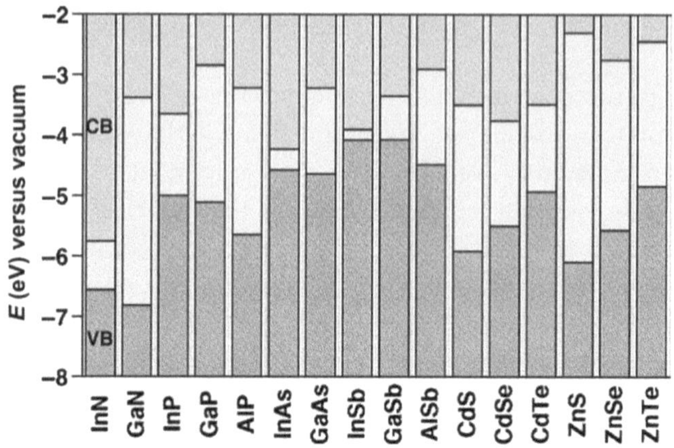

Figure 2.20 Electronic energy levels of some III–V and II–VI semiconductor materials, CB indicate the conduction band, while VB indicates the valence band edges.

that can be achieved due to different band alignment. Hence the bandgap, conduction band, and valence band edges positions should carefully be chosen before deciding on the CSNP material candidates. Fig. 2.20 shows some selected III–V and II–VI semiconductor band positions [46] (Fig. 2.21).

By using an appropriate band edge alignment for both the core and shell material a specific type of the heterojunction interface of the CSNP can be obtained. Depending on the bandgap and the relative position of the valence and conduction band edges positions, the shell will have a specific function [46]. As can be seen there are, in principle, three different types of possible band alignment realized so far. These are type I, reverse type I, and type II. In type I, the bandgap of the shell is larger than that of the core material leading to confinement of both electrons and holes in the core material. In the reverse type I, the bandgap of the shell is smaller than that of the core. So depending on the thickness of the shell, the charged carriers will be completely or partially confined in the shell [46]. In type II, either the conduction band or valence band of the shell are located within the bandgap of the core material. Because of this, and when the system is excited, the holes and electrons will be separated to reside on different parts of the core–shell system, that is, each carrier will either be located in the core or in the shell while the other carrier will be on the other side. So in type II, charge separation will prevail upon external excitation [46]. The shell in the three different types is introduced to perform different functions in each case. For type I

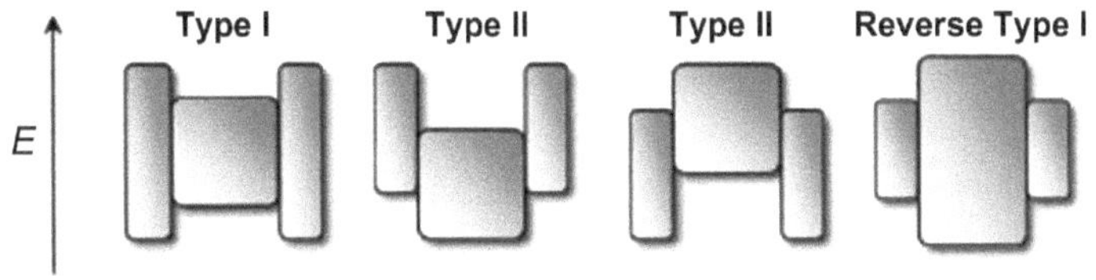

Figure 2.21 Schematic representation of the energy band alignments in different core–shell nanoparticles. The upper and lower edges of the outer rectangles represent the conduction and valence band edges, respectively, of the shell material, while those of the square in the middle represent the conduction and valence band edges, respectively, of the core material.

CSNPs, the shell is introduced to act as a physical shield for the optically active core. Here the shell physical separates the optically active core from molecules of the surrounding environment, like, for example, oxygen or water molecules. Such CSNPs possess stability against photodegradation. Beside the function of physically separating the core, in type I CSNPs the shell also reduces the number of surface dangling bonds at the surface of the core. Since these surface dangling bonds can act as trapping centers leading to reductions of the quantum yield, the shell can lead to improvement of the optical properties of the optically active core. The first demonstrated CSNP of type I is the CdSe/ZnS system. The coverage of the CdSe with a ZnS shell leads to improving the quantum yield of the CdSe. The optical response of the CdSe is observed with a relative redshift of about 5–10 nm [46]. In reverse type I, another narrower bandgap shell material is grown around a wider bandgap core material. Here almost all the carriers will be confined in the shell. Here a relatively large redshift is observed. The magnitude of the redshift will depend on the shell thickness. The most studied reverse type I systems are CdS/HgS, CdS/CdSe, and ZnSe/CdSe [49–51]. To improve the photobleaching and the quantum yield, a second shell of a material with a wider bandgap can be overgrown on the first shell [46,50]. In type II CSNPs, the shell function is primarily to induce a relatively large redshift of the absorption or emission spectra. The formation of a staggered bandgap alignment in a type II system leads to a smaller effective bandgap of the system compared to the two materials in the system. This system, that is, type II CSNP, is usually utilized to tune the emission color, something that is not easily achieved by other materials and systems. The most studied type II systems are CdTe/CdSe and CdSe/ZnTe [46]. For type II CSNPs the photoluminescence spectra decay times are relatively longer compared to those observed from type I CSNPs [46]. The reason for this is the lower overlap of the electrons and holes wave functions. Since one of the charged carriers, that is, electrons or holes, resides in the shell, and the other resides in the core, the overgrowth with another second wider bandgap shell can lead to improving the optical quantum yield and increasing the photostability [46]. As mentioned above, the addition of a second shell in CSNPs can lead to many benefits. Among them is the increase of the quantum yield. Nevertheless, sometimes and due to the lattice mismatch between the crystal parameters of the core and shell, it is essential to introduce intermediate material is essential to reduce the relative lattice mismatch. It is important to mention that the lattice mismatch between the shell and the core can lead to the introduction of interface defects which will deteriorate the optical properties and lead to relatively low quantum yield. Hence in some cases the induction of strain-releasing intermediate material between the core and the outer shell is unavoidable. Such systems are called core/shell/shell nanoparticles (CSSNPs). Fig. 2.22 illustrates the different types of CSSNPs with different bandgaps and band alignments [46].

The first demonstrated CSSNP system with an intermediate strain-releasing layer was CdSe/ZnSe/ZnS [52]. The band alignment structure of such a system is shown to the far left of Fig. 2.22. Such CSSNPs are shown to have a better quantum yield, possess improved photodegradation, and have a better charge confinement [53]. Also the intermediate layer can act as a barrier for carriers and hence increase the radiative life time. For CdSe/CdTe/ZnTe the carriers radiative life time is as long as 10 ms, this is because the intermediate CdTe layer acts as a barrier for electrons (residing in the CdSe) and hole (residing in the ZnTe) [54].

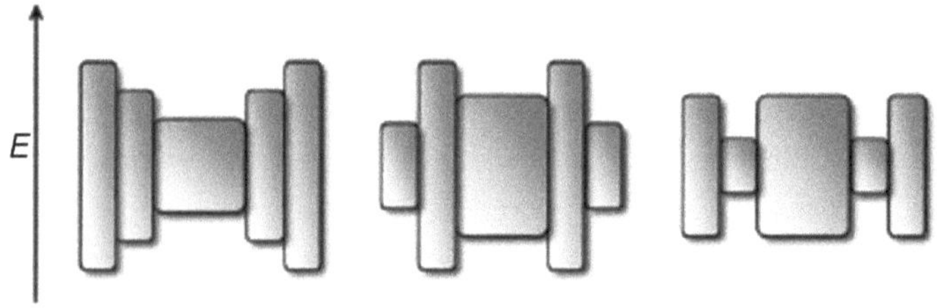

Figure 2.22 Schematic presentation of the energy band alignment in different CSSNPs. The height of the rectangles represent the bandgap energy and their upper and lower edges correspond to the position of the conduction and valence band edges, respectively, of the core (in the center) and the two shell materials. *CSSNPs*, Core/shell/shell nanoparticles.

A special case of the CSSNP system is the so-called quantum dot quantum well structures (QDQWs). Here a lower bandgap material is embedded between two larger bandgap materials (a core and the outer shell). This is illustrated in the right part of Fig. 2.22. Among the first demonstrated systems of QDQW structure was CdS/HgS/CdS [55]. Such a QDQW system can be synthesized by immersing the CdS/HgS in H_2S, and due to the difference in solubility of CdS and HgS, Cd will replace some of the Hg and transform a thin layer of the HgS into CdS. This method will be discussed in detail in an upcoming chapter as it is a low-temperature bottom-up method with potential for many applications.

An inverse QDQW system, where a large bandgap material, that is, the inner shell, is impeded between two smaller bandgap materials (the core and the outer shell), has also been demonstrated with interesting optical properties. This is shown in the middle schematic representation in Fig. 2.22. CdSe/ZnS/CdSe is a typical example of an inverse QDQW CSSNP system [56]. This reverse CdSe/ZnS/CdSe QDQW system represents the first heterostructure that simultaneously provides two emission wavelengths coming from the outer shell as well as from the inner core, both from CdSe material. For the decoupling between the core and the outer shell the intermediate ZnS has to have a minimum thickness of about three monolayers [56]. The light harvesting from this CdSe/ZnS/CdSe CSSNP has been studied in detail in [57]. If in this structure system we consider the inner CdSe is a core, while the ZnS is a barrier and the outer CdSe is a shell, the following scenario is suggested: When the CdSe shell absorbs photons, a consequent emission from the CdSe core materials is observed. The quantum well CdSe shell acts as a light harvester, which increases the brightness of the CdSe core [58]. It is then shown that both the CdSe core and shell are coupled by excitons tunneling through the ZnS intermediate barrier [58]. As white light is a mixture of three different wavelengths compromising red, green, and blue wavelengths, the simultaneous emission depicted in Fig. 2.23 above, can be utilized to develop white light sources. Indeed this has been demonstrated by extending the CdSe/ZnS/CdSe QDQW heterostructure to include an extra ZnS third shell [57], that is, forming a CdSe/ZnS/CdSe/ZnS core–triple shell system triple core–shell (c3s) system [57].

Fig. 2.24 shows the photoluminescence spectra of the as-grown and annealed CdSe/ZnS/CdSe/ZnS core–triple shell (c3s) NPs. As discussed above it was shown that the CdSe/ZnS/CdSe could be designed to decouple the light emission from the core and the outer shell by

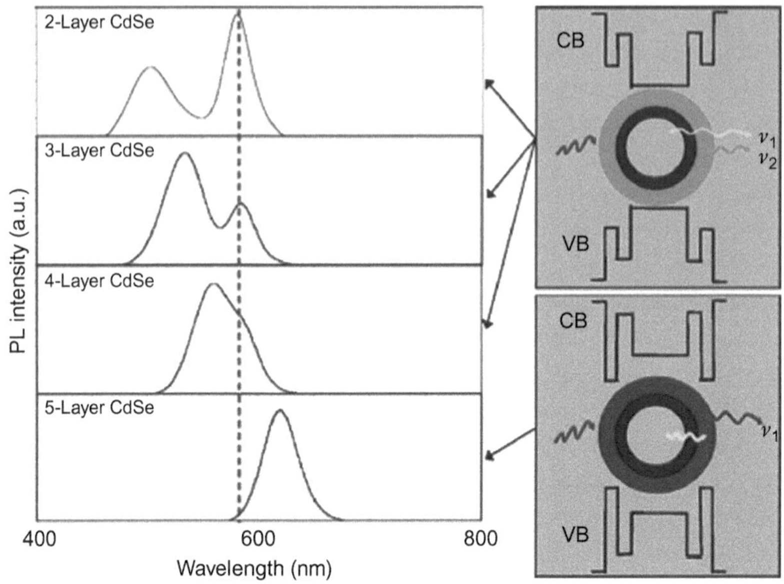

Figure 2.23 CdSe/ZnS/CdSe CSSNP system. The left side shows photoluminescence spectra as a function of the CdSe out shell thickness. To the right the corresponding energy band diagram displays different absorption and emission processes [56]. *CSSNPs*, Core/shell/shell nanoparticles.

introducing an intermediate wider bandgap material, ZnS in this case [56]. Utilizing this decoupling, and as is clear from Fig. 2.24, it is possible to adjust the ratio of the different colors to obtain white light emission if a third shell of ZnS is grown on the second CdSe shell [56,57]. In Fig. 2.24 exciting the c3s structure by 400 nm wavelength, the as-grown sample emitted an orange color. While annealing at 200°C for 50 minutes the emitted color turns white [57]. Upon heating for longer durations, the blue component of the white color is increased and the emitted color is no longer white. This implies that a careful design of such a c3s structure combined with post-growth annealing can be used to obtain intrinsic white light [57]. Also it worth mentioning that in Fig. 2.24, the lower spectrum is for trap-rich CdSe NPs. Although in the same experiments, sometimes trap-rich CdSe NPs have emitted white light, and sometimes they do not. As can be seen, the trap-rich CdSe NPs emit only orange color with no blue component and hence there is no chance of obtaining white light from CdSe NPs alone [57]. This is an example of the effect of morphology of small-size structures in tuning the optical properties.

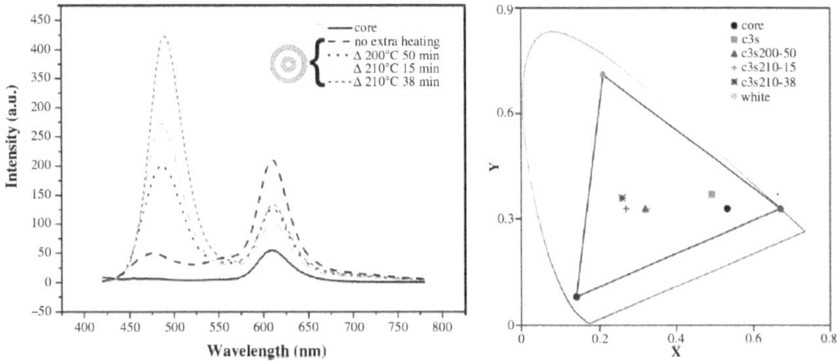

Figure 2.24 Left: Photoluminescence spectra of as-grown and annealed CdSe/ZnS/CdSe/ZnS core–triple shell system. Right: the corresponding CIE coordinates of the emitted light. Here c3s denotes the core–triple shell system.

2.5 Nanodielectric effects

So far, we have discussed the effects of nanosized materials for semiconductors as well as metals. Nevertheless, dielectrics, which is a third class of materials, have also pronounced effects when scaled down to the nanometer size. Below we discuss the effects and benefits that can be gained when a dielectric material is scaled down to the nanometer size. Lewis [59] first introduced the topic of nanodielectrics in 1994. In this pioneer ing work, Lewis predicted that designing and implementing dielectrics at the nanoscale can lead to very useful properties of dielectric and conductive interfaces. Such nanoscale interfaces will have molecular organization with noncentro symmetry as an inherent characteristic of such nanodielectric interfaces. Such interfaces have potential for improved sensors, actuators, and transducer devices [59]. The experimental evidence of the benefits of electrical insulation using nanodielectrics was shown shortly after Lewis' pioneering work by Henk et al. and others [60,61]. In a study, titanium dioxide (TiO_2) nanoparticles (23 nm) were incorporated into an epoxy matrix to form a nanocomposite. It was shown that the properties of the nanocomposite were tremendously improved. Electric strength short-term measurements under direct current (DC) conditions with an amp of 500 V/s have been performed for the composite of the epoxy with TiO_2 nanoparticles with different sizes (23 nm and 1.5 μm sizes) [61]. The results are shown in Fig. 2.25. As can be seen, the breakdown gradient for micro- and nanosized composites as a function of the filter loading indicate very clearly the benefits that can be gained from the

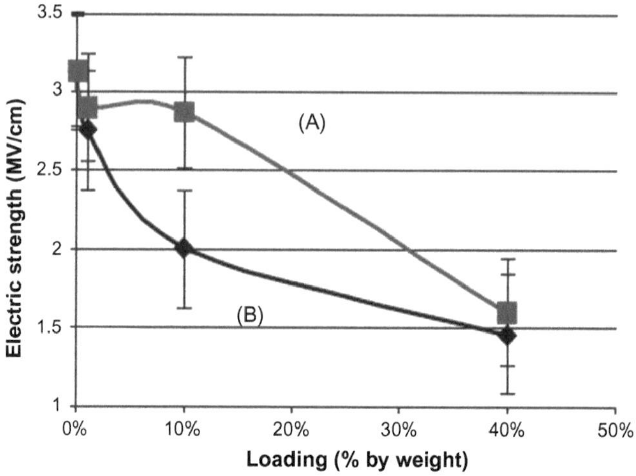

Figure 2.25 Electrical strength of Epoxy/TiO_2 composites. (A) 23 nm, and (B) 1.5 μm [61].

use of a nanometer-size nanoparticle to form the composite. The conclusion that has been drawn from this study to explain the improvements is that when the size of the incorporated nanoparticles become small enough, they act cooperatively with the host matrix and are no longer acting as guests forming an interface. This will lead to Maxwell–Wagner polarization [61]. It is further postulated that the small-size inserted particles are surrounded by high charge concentrations in a Gouy–Chapman–Sterner layer. Due to the large specific surface area of these relatively small-size inserted nanoparticles, limited charge percolation will be allowed, and hence the electrical insulation properties are enhanced by the incorporation of such nanometer-sized nanoparticles in the insulator host matrix [61].

References

[1] S.V. Boriskina, J.K. Tong, W.-C. Hsu, L. Weinstein, X. Huang, J. Loomis, et al., Proc. SPIE 9546, Active Photonic Materials VII, (20159, 95461U); <https://doi.org/10.1117/12.2189679>.
[2] M.J. Madou, R. Cubiccoiotti, Proc. IEEE 91 (2003) 830.
[3] C.N.R. Rao, P.J. Thomas, G.U. Kulkarni, Nanocrystals: Synthesis, Properties, and Applications, Springer, 2007.
[4] A. Henglein, Chem. Rev. 89 (1989) 1861.
[5] G. Schön, U. Simon, Colloid Polym. Sci. 273 (1995) 101.
[6] U. Simon, G. Schön, G. Schmid, Angew. Chem. Int. Ed. Engl. 232 (1993) 250.
[7] Silicon or any other semiconductor band structure.
[8] M. Freitag, Nat. Phys. 7 (2011) 596–597.
[9] M. Kastner, Phys. Today 46 (1993) 24.

[10] L.P. Kouwenhoven, C.M. Marcus, R.M. Westervelt, N.S. Wingreen, Electorn transport in quantum dots, in: NATO ASI Series Volume 345 of the series, 1997, pp. 105–214.
[11] H. Grabert, M.H. Devoret (Eds.), Single Charge Tunneling, Plenum Press, 1991.
[12] D.V. Averin, K.K. Likharev, J. Low Temp. Phys. 62 (1986) 345; and in: Mesoscopic Phenomena in Solids, B.L. Altshuler, P.A. Lee, R.A. Webb (Eds.), Elsevier, 1991.
[13] X. Luo, A.O. Orlov, G.L. Snider, Microelectron. J. 36 (2005) 308–312.
[14] A.G. de Aguila, (Ph. D. thesis), Radboud Unversity/Universiteit Nijmegen, The Netherlands, 2015.
[15] F. Furis, H. Htoon, M.A. Petruska, V.I. Klimov, T. Barrick, S.A. Crooker, Phys. Rev. B 73 (2006) 241313.
[16] A.M. Smith, S. Nie, Acc. Chem. Res. 43 (2) (2010) 190–200.
[17] N.T. Bagraev, N.G. Galkin, W. Gehlhoff, L.E. Klyachkin, A.M. Malyarenko, J. Phys.: Condens. Matter 20 (2008) 164202.
[18] B.J. van Wees, H. van Houten, C.W.J. Beenakker, J.G. Williamson, L.P. Kouwenhoven, D. van der Marel, et al., Phys. Rev. Lett. 60 (1988) 848. Phys. Rev. B 43, 12431 (1991).
[19] R. Landauer, IBM J. Res. Dev. 1 (1957) 223.
[20] Yu. V. Sharvin, Sov. Phys. JETP 21 (1965) 655.
[21] K.L. Kelly, E. Cononado, L.L. Zhao, G.C. Schatz, J. Phys. Chem. B 107 (2003) 668.
[22] G. Mie, Am. Phys. 25 (1908) 377.
[23] P.F. Liao, M.B. Stem, Opt. Lett. 7 (1982) 483.
[24] C.P. Collier, R.J. Saykally, J.J. Shiang, S.E. Henrichs, J.R. Heath, Science 277 (1997) 1978.
[25] R. Jin, Y. Cao, C.A. Mirkin, K.L. Kelly, G.C. Schatz, J.-G. Zheng, Science 294 (2001) 1901.
[26] C.K. Chen, A.R.B. de Castro, Y.R. Shen, Phys. Rev. Lett. 46 (1981) 145.
[27] H. Metiu, P. Das, Annu. Rev. Phys. Chem. 35 (1984) 507.
[28] J.I. Dadap, J. Shan, K.B. Eisenthal, T.F. Heinz, Phys. Rev. Lett. 83 (1999) 4045.
[29] W.H. Yang, J. Hulteen, G.C. Schatz, R.J. Van Duyne, Chem. Phys. 104 (1996) 4313.
[30] E. Hao, G.C. Schatz, J.T. Hupp, J. Fluoresc. 14 (2004) 331.
[31] W.H. Yang, G.C. Schatz, R.P. Van Duyne, J. Chem. Phys. 103 (1995) 869.
[32] T. Jensen, K.L. Kelly, A. Lazarides, G.C. Schatz, J. Cluster Sci. 10 (1999) 295–317.
[33] R.X. Bian, R.C. Dunn, X.S. Xie, P.T. Leung, Phys. Rev. Lett. 75 (1995) 4772–4775.
[34] K.L. Kelly, A.A. Lazarides, G.C. Schatz, Comput. Sci. Eng. 3 (2001) 67.
[35] S. Link, M.A. El-Sayed, J. Phys. Chem. B 103 (1999) 8410.
[36] M. Treguer, C. de Cointet, H. Remita, J. Khatouri, M. Mostafavi, J. Amblard, et al., Phys. Chem. B 102 (1998) 4310.
[37] S. Link, Z.L. Wang, M.A. El-Sayed, J. Phys. Chem. B 103 (1999) 3529.
[38] M.C. Lea, Am. J. Sci. 37 (1889) 476.
[39] P. Mulvaney, M. Giersig, A. Henglein, J. Phys. Chem. 97 (1993) 7061.
[40] G.C. Papavassiliou, Prog. Solid State Chem. 12 (1980) 185.
[41] R. Gans, Ann. Phys. 47 (1915) 270.
[42] P.B. Johnson, R.W. Christy, Phys. Rev. B 6 (1972) 4370 (1972).
[43] S. Link, M.B. Mohamed, M.A. El-Sayed, J. Phys. Chem. B 103 (1999) 3073.
[44] Y. Yu, S. Chang, C. Lee, C.R.C. Wang, J. Phys. Chem. B 101 (34) (1997) 6661.
[45] B.M.I. Zande, M.R. Bohmer, L.G.J. Fokkink, C.J. Schonenberger, Phys. Chem. B 101 (1997) 852.
[46] P. Reiss, M. Protiere, L. Li, Small 5 (2009) 154.
[47] A. Sadollahkhani, I. Kazeminezhad, J. Lu, O. Nur, L. Hultman, M. Willander, RSC Adv. 4 (2014) 36490.

[48] A. Sadollahkhani, O. Nur, M. Willander, I. Kazeminezhad, V. Khranovskyy, M.O. Eriksson, et al., Ceram. Int. 40 (2015) 7174.
[49] A. Mews, A. Eychmuller, M. Giersig, D. Schooss, H. Weller, J. Phys. Chem. 98 (1994) 934.
[50] D. Battaglia, J.J. Li, Y.J. Wang, X.G. Peng, Angew. Chem. Int. Ed. 42 (2003) 5035.
[51] X.H. Zhong, R.G. Xie, Y. Zhang, T. Basche, W. Knoll, Chem. Mater. 17 (2005) 4038.
[52] J. Bleuse, S. Carayon, P. Reiss, Phys. E 21 (2004) 331.
[53] D.V. Talapin, I. Mekis, S. Gotzinger, A. Kornowski, O. Benson, H. Weller, J. Phys. Chem. C 112 (2008) 1744.
[54] C.Y. Chen, C.T. Cheng, C.W. Lai, Y.h. Hu, P.T. Chou, H.T. Chiu, Small 1 (1215) 205.
[55] A. Eychmuller, A. Mews, H. Weller, Chem. Phys. Lett. 208 (2001) 59.
[56] D. Battaglia, B. Blackman, X.G. Peng, J. Am. Chem. Soc. 127 (2005) 10889.
[57] S. Sapra, S. Mayilo, T.A. Klar, A.L. Rogach, J. Feldmann, Adv. Mater. 19 (2007) 569.
[58] E.A. Dias, S.L. Sewall, P. Kambhampti, J. Phys. Chem. C 111 (2007) 708.
[59] T.J. Lewis, IEEE Trans. Dielectr. Electr. Insul. 5 (1994) 812.
[60] P.O. Henk, T.W. Kortsen, T. Kvarts, High Perf. Poly. 11 (1999) 281.
[61] J.K. Nelson, J.C. Fothergill, Nanotechnology 15 (2004) 586.

Further reading

Celso de Mello Donega, Chem. Soc. Rev. 40 (2011) 1512–1546.
M. Nirmal, C.B. Murray, M.G. Bawendi, Phys. Rev. B 50 (1994) 2293.

CHAPTER 3

Conventional nanofabrication methods

3.1 Introduction

Although the driving force that led to the era of nanotechnology was mainly electronics, and that was necessitated by the need for faster and handier personal electronics, nowadays, the nanotechnology bandwagon is rolling faster and faster including new applications every now and then. This is owing to the fact that nanomaterials provide an amazingly wide range of unusual properties. Hence it is obvious that the methods and techniques of electronics fabrication, that is, lithography etc., are among the first modern conventional techniques human being used to produce nanomaterials. This is despite the fact that much older primitive techniques were used before the era of modern electronics, for example, mechanical grinding, for the production of nanoparticles (NPs). In general the synthesis of nanostructures/nanomaterials can be achieved by four different generic routes. These are the mechanical approach, the wet chemical routes, the gas phase route, and the form-in-plane approach [1]. All these routes, except the wet chemical approach, which is the main topic of this book, will be discussed in this chapter. Despite the progress in improving existing techniques and/or developing new techniques for nanofabrication, some important issues are to be highlighted and resolved before nanomaterials can be fully utilized. The issues to be studied investigate the effect of nanomaterials on the environment and health as well as the ethical aspect of nanomaterials [1]. Before reaching consensus on the safety of nanomaterials, the industry will not risk any huge investments.

As this chapter is titled "conventional nanofabrication methods," we will concentrate on the conventional techniques used to realize nanomaterials. Priority will be devoted to the most commonly used techniques. Both conventional top-down and bottom-up approaches will be introduced and presented, with some examples of their use for fabricating nanostructures for different applications.

Low Temperature Chemical Nanofabrication
DOI: https://doi.org/10.1016/B978-0-12-813345-3.00003-4

3.2 Mechanical techniques

Mechanical methods include grinding, milling, and mechanical alloying. As long as the product appears as a coarse powder of targeted dimension. In principle, mechanical methods rely on the old-age technique of physical crushing a bulky material into smaller and smaller units, down to the nanometer dimensions. Usually the required structure is the same as the starting bulky material. No chemical transformation is usually observed after the grinding/milling. Although these mechanical approaches have many advantages, like the simplicity, low cost, etc., they suffer from many disadvantages. The main disadvantages are contamination and agglomeration of the produced powder, leading to wide size distribution. This is in addition to the difficulty in achieving relatively smaller sizes, and relatively long operation hours, in access of 100 hours, in some cases.

Fig. 3.1A shows a typical high-energy miller part (A), while Fig. 3.1B shows a horizontal cross section of the miller together with the direction of rotation of both the miller disk and the balls in the bowl. The most common mechanical approach for producing nanomaterials is ball milling. This technique was discovered at the end of the 1960s. The technique is termed mechanical alloying and it was found to produce rather fine oxide particles (Al_2O_3, Y_2O_3, ThO_2) in nickel-base super alloys that it was not possible to produce using other conventional powder metallurgy methods [2]. There are different apparatus used in this technique, but the principle of operation is the same in all the used apparatus. The most common apparatus is the planetary ion millers. Usually these planetary millers can

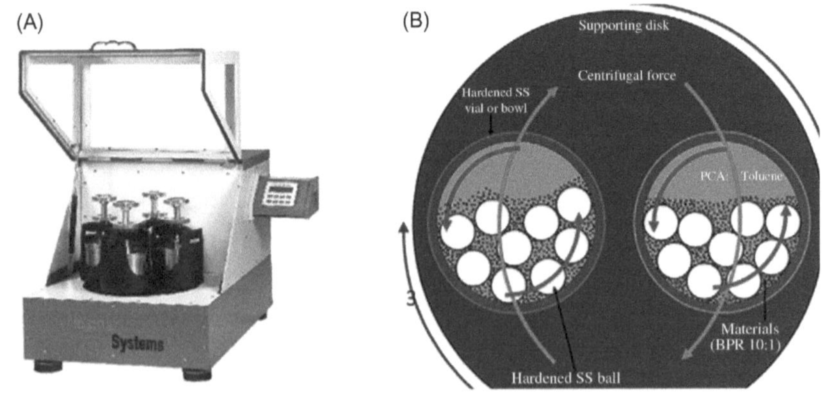

Figure 3.1 (A) Typical mechanical fabrication machine used of nanomaterials. (B) Schematic diagram of the mechanical fabrication of nanostructures.

operate with a small amount of powder and hence are suitable for research laboratories. This system consists of one turn disk and two to four bowls. The bowls rotate in one direction and the turn disk rotates in the opposite direction (see Fig. 3.1B). These two parts rotating in opposite directions act on the powder and lead to fracturing of the powder under the experienced high-energy impact. The powder in an operating high-energy planetary milling apparatus experiences an energetic impact due to the rotation of the two main parts of the miller. The whole grinding process can be divided into four different parts. The four different parts are [3]:

1. Initial stage: Here the particles are first flattened due to the collision of the balls. Microforging will then leads to the modification of the shape of the initial particles as a result of the collision with high kinetic energy mill balls. Nevertheless, at such stage the net mass of the powder will not change.
2. Intermediate stage: At this stage, cold welding will be significant. The mixture of the powders will possess a decreased diffusion distance down to the micrometer region. It is to be noted that although some dissolution happens, the chemical composition of the powder is still not homogenous.
3. Final stage: At this stage, reduction in the particles size will be achieved. Further, a microscopic scale indicates that the powder is more homogenous compared to the two first stages.
4. Completion stage: At this stage the powdered particles will possess extremely deformed metastable configuration and the lamellae cannot be resolved by optical microscopy. No further improvement on the dispersion of the produced relatively small size powder can be achieved beyond this stage. The final product will hence be a real alloyed small size powder similar to the starting materials.

Varying the parameters of the alloyed materials in milled nanostructure can be a way to tune and modify the composite size. Fig. 3.2 shows the variation of the size of an alloyed Al and TiO_2 as the original TiO_2 percentage is varied [3].

Although the applications of mechanochemically synthesized nanomaterials are rather limited, the technique is unique for certain classes of materials needed for specific applications. A typical example is that metastable compounds with a local nonequilibrium structure can be difficult to obtain by utilizing known solid-state preparation techniques that operate at relatively elevated temperatures. Mechanical treatment of two or three starting materials in high-energy ball mills to form a composite

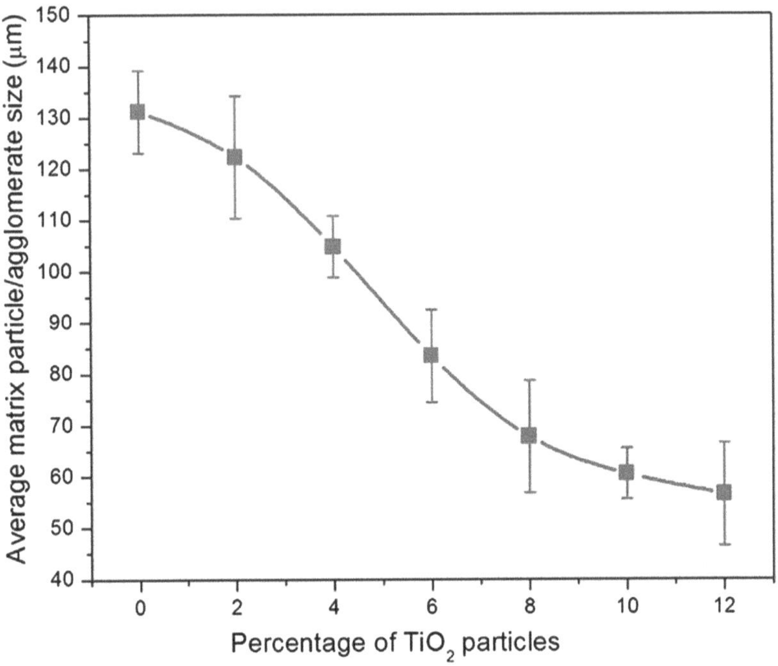

Figure 3.2 The average particle size dependence on the percentage of theTiO_2 particles.

enabling the synthesis of not only new, metastable compounds but also of nanocrystalline materials with unusual or enhanced properties.

Hence nanocomposites prepared using such mechanochemical techniques have found their uses in many applications, for example, dyes, hardened materials for automobile applications. More recently the interest has even been extended to some applications in electronics. Due to its elemental abundance, sodium (Na) is considered, at the moment, to be the best replacement for Lithium ion batteries [4]. This fact is clearly seen in sodium-based ion batteries (NIB) research activity. In NIB similar structural types of sodium-based electrodes compromise the cell elements [4]. Antimony (Sb) has been suggested as the element to be combined with Na to form NaSb. But due to its relatively large mass, scarcity, and toxicity, Sb is not the best candidate [5]. Using the high-energy planetary ball milling technique, different Na nanocomposite compounds were demonstrated and characterized [4]. In these experiments, encouraging results for the mechanochemical room temperature synthesis of Na compounds indicate the possible potential of this technique. This is due to the fact that it

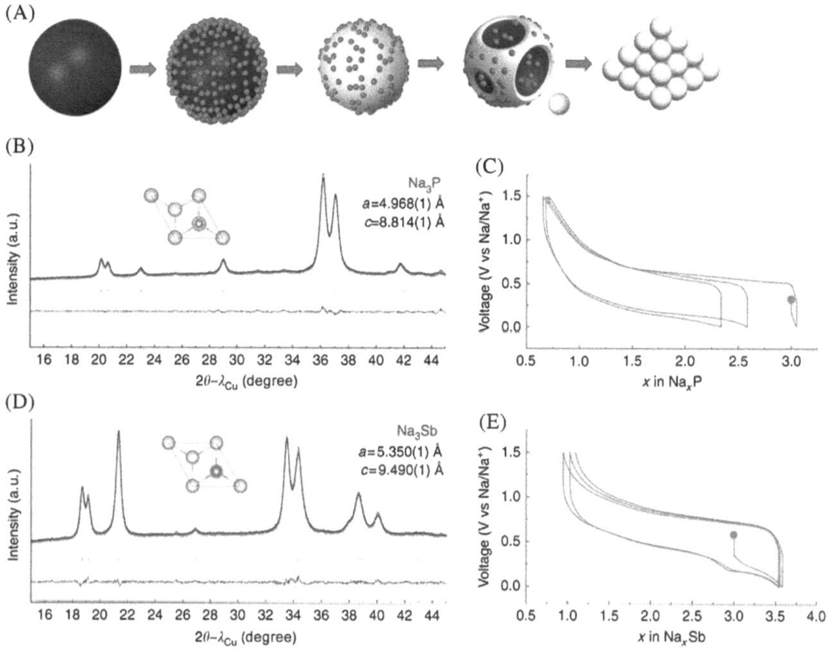

Figure 3.3 (A) Schematic diagram showing the ball milling process exemplified for the formation of NaxM alloys by reacting Na with metal M (M = P or Sb). In (B and C) and (D and E) the Rietveld refined X-ray powder patterns of Na_3P and Na_3Sb powders are, respectively, shown together with their corresponding electrochemical voltage profile in Na-half cells.

can be scaled up for industrial mass production. Although a technique can be very suitable for producing nanomaterial for research purposes, if this technique is not possible for scaling up the produced material then it is not of interest. Fig. 3.3 shows a simplified schematic diagram of the process, accompanied by different characterization results [4]. Despite this success the fact that the produced nanomaterials using this technique and the fact that the product is freestanding limit the possible applications of the product. Further it is to be noted that this route of obtaining nanomaterials is a low-temperature process, but it is not widely utilized due to the fact that the produced material is freestanding.

3.3 Lithography techniques

Lithography was invented in 1798 by Alois Senefelder and initially based on the use of Gum Arabic as the material to transfer a specific pattern and limestone as a substrate. Later and specifically in 1950 limestone was replaced by silicon and Gum Arabic was replaced by resist. There are many top-down methods used for nanomaterial fabrication used in the electronics industry and they are all collectively called lithography. In principle all these methods use light, an electron beam, or an ion beam to transfer a pattern to a substrate. To achieve this transfer a precursor material, called the resist, is used. As the electronics industry is advancing very rapidly, even faster than Moore's law, the minimum feature size in the pattern transfer process has been the target of reduction continuously. Fig. 3.4 shows a schematic diagram illustrating the principle of lithography. After transferring the pattern to the substrate that is covered by the resist, etching, metal deposition, or doping by implantation occurs [6].

As shown in Fig. 3.4, there are two types of resist. The negative and positive resist and both are photosensitive materials. The negative resist material bonds break when irradiated with photons, ions, or electrons, while in the positive resist the bonds become stronger. Hence after irradiation with a suitable calibrated dose of photons, ions, or electrons, the sample is developed in a special solution to remove the irradiated part (negative resist) or nonirradiated part (for the positive resist). Working with

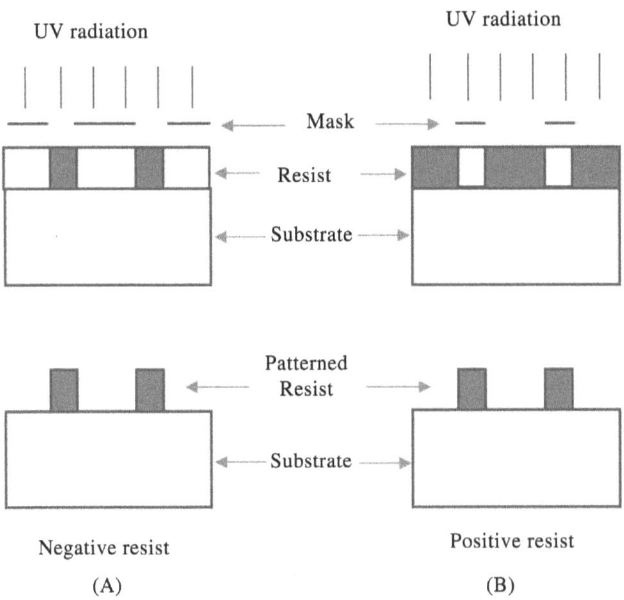

Figure 3.4 Illustration of the first step in lithography when using UV radiation: (A) for the case of a negative resist, while in (B) for the case of positive resist. *UV*, Ultraviolet.

negative resist is much easier and hence it is commonly used in fabricating integrated circuits (ICs). In general, the resolution (R) of the lithography technique, that is, the minimum feature size that can be achieved, is given by:

$$R = k\frac{\lambda}{NA} \tag{3.1}$$

where λ is the wave length of the radiation used, k is a constant, and NA is the numerical aperture. The constant k is process dependent—usually in IC manufacturing the value of k lies between 0.5 and 0.8. The value of NA is rather close to k. Hence as a rule of the thumb, the resolution R is taken usually to be equal to the wavelength of the radiation used in the specific lithography system [6]. In recent decades there has been tremendous improvement of the minimum feature size possible to pattern using lithography. This improvement is driven by the demand for reduced size electronic devices and equipment.

There are many top-down techniques used for the fabrication of nanostructures, however, here we will briefly discuss the most commonly used lithography-based nanofabrication techniques. Although lithography is based on pattern transfer, it is divided into two main types. These are:

1. Physical mask methods or mask lithography: In these methods the photosensitive material, that is, the resist, is irradiated through a mask that is in contact or proximity with the surface of the sample.
2. Software methods or scanning methods: Here the pattern is sequentially transferred by a direct beam irradiated on the sample through a computer controlled software. It is in principle a transfer of a mask drawn on a computer program.

The main differences between the mask and scanning methods are the speed and resolution. The speed of the scanning methods is much less than the mask method. This is because the mask method is a parallel technique while the scanning methods are a serial technique. Also the resolution of the scanning methods is much better and smaller feature sizes can be fabricated. The difference in resolution is basically due to the irradiation source used. In case of the scanning much shorter irradiation wavelengths can be utilized. Another important difference. Below we will describe the most commonly used lithography techniques.

3.3.1 Optical lithography

This technique uses light as the radiation source. Ultraviolet (UV) of different types (UV, deep UV, extreme UV) and X-ray are the usual sources for irradiation. These irradiation sources are used together with a mask. The mask is made of glass or quartz and the light sensitive layer is exposed to the radiation. During this exposure, a specific pattern is transferred. After a calibrated exposure, the light sensitive layer (photoresist) is developed and the mask pattern is then achieved on the substrate, as shown in Fig. 3.4. This pattern transfer in photolithography is achieved through two different procedures: either by direct contact between the mask and the substrate—his is called contact lithography; or projection lithography where the pattern passes through lenses which reduce the size of the pattern projected on the substrate. This is called projection lithography. Usually in projection mode lithography the pattern dimensions are reduced by a factor of 5–10 times. In optical lithography the minimum feature size is in the range of 0.5–0.8 μm when using 360–460 μm UV radiation. Better resolution cannot be achieved due to the difficulty in reducing the gap between the substrate and the mask, which is usually about 1 μm as a vacuum is applied in the system. To obtain lower feature sizes down to or below 100 nm scale extreme UV and X-ray radiation is employed. However, the cost of such systems is extremely high compared with conventional optical photolithography, due to the high cost of the equipment using extreme UV and X-ray radiation.

3.3.2 Scanning mode lithography

As mentioned above, the rule of the thumb of the minimum feature size possible to obtain using lithography is that this minimum feature size is equal to the wavelength of the used radiation. Hence, to obtain feature sizes in the nanometer regime (1–100 nm) a radiation with a suitable wavelength should be used. Energetic particles (electrons and ions) can provide such a possibility. When using electrons the technique is called electron beam lithography (EBL), while when using ions the techniques are called collectively focused ion beam lithography. Such top-down techniques are suitable for nanodevices fabrication and they are routinely used in many research laboratories. It is important to mention that these techniques, that is, scanning mode lithography, have found interesting applications not only for electronic devices, but also in other fields, like biology etc. Such applications have added the dimension of interdisciplinary nature to the field of nanotechnology.

3.3.3 Electron beam lithography

EBL started to appear in the late 1960s, and although sizes down to 50 nm are routinely demonstrated, nowadays, this minimum feature size has been reduced down to about 5 nm in principle using advanced EBL equipment. Hence the EBL is considered as one of the techniques with the highest resolution. The EBL is quite expensive and complicated, and very few companies produce and develop EBL systems. The EBL technique is similar in principle to conventional lithography, that is, it is based on three steps. These are the exposure of the sample surface, followed by developing the exposed area, and finally the pattern transfer. Both positive and negative photoelectron sensitive resists are used. A typical example of an electron sensitive material that is routinely used is poly(methyl methacrylate) denoted as PMMA. Fig. 3.5 shows a typical EBL system.

To achieve high resolution, minimum electron scattering in the EBL resist is essential. An EBL system is composed of mainly four basic parts. These are the electron source gun (produces the beam of electrons with a specific energy and a specific dose), electron column (this part forms and focuses the electron beam), mechanical stage (with the function to accurately locate the beam), and finally a control computer with a special software. As mentioned above, to achieve precise patterns with high resolution, the electron sensitive resist is usually composed of a thin layer. Nevertheless, this thin layer feature creates a difficulty in processing standard steps. Mainly the lift-off of any deposited layer after the developing

Figure 3.5 A typical electron beam lithography system. *(From Raith GmbH.)*

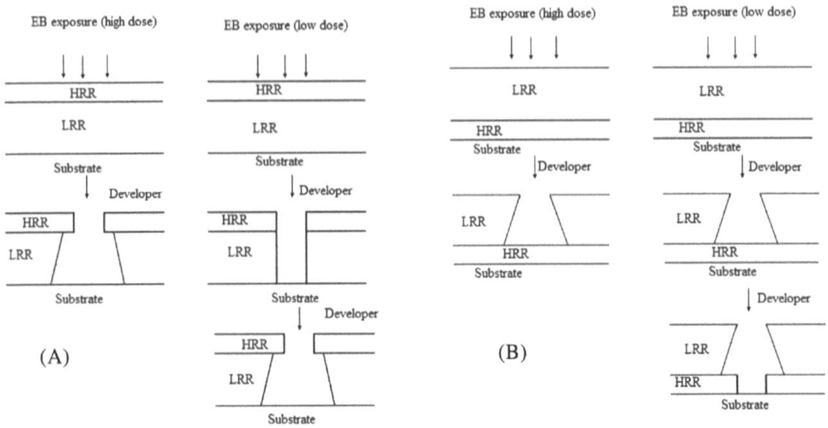

Figure 3.6 Schematic diagram of the double layer resist HRR and LRR for lift-off technique using electron beam lithography in (A) HRR on top of a LRR and (B) LRR on top of a HRR. *HRR*, High-resolution resist; *LRR*, low-resolution resist.

will be difficult. For this purpose, a double layer configuration was developed to overcome the lift-off difficulty [7]. This double layer resist is composed of a high-resolution (low scattering) top thin resist with a lower resolution (high scattering) thicker resist. Also, depending on the pattern required, the sequence of the layers can be reversed. A typical example of a double layer configuration is shown in Fig. 3.6. As can be seen the thicker low-resolution resist will suffer from the so-called undercut which enables an easy lift-off. To reduce the scattering, particles with a mass higher than electrons are sometimes used. Typically, H^+, He^{++}, Li^+, and Be^{++} ions are the typical ions used when replacing the energetic electron beam. There are two basic types of EBL. The first is the scanning EBL system. In this system, a focused beam of electrons exposes the resist by moving the pane of the pattern. Here in this technique there is no mask or template required. The disadvantage is that it is a sequential scanning and hence it requires long exposure durations and this makes the process relatively slow.

The second is projection electron beam lithography. Here a relatively wide unfocused beam of electrons is used to expose the whole pattern through one exposure. In this system a photocathode is placed over an optical mask that has a specific pattern. When UV radiation is focused through the mask, emission of electrons from ta specific area in the photocathode will take place. Then by using static and magnetic fields, the beam of electrons having the shape of the pattern will be exposed on the sample surface in one exposure. This makes the projection EBL faster than the scanning EBL method. Nevertheless, all EBL techniques are serial techniques and hence are very slow, which restricts the use of EBL applications. In addition, a typical EBL system is relatively expensive. This makes the application of EBL restricted to mask making and for research purposes in developing prototype devices and patterns.

It is to be noted that the EBL technique can also be operated in low-resolution high-speed and high-resolution low-speed modes. The design of a chip when using EBL should include global alignment marks and local alignment marks. The global alignment marks are used to locate the whole substrate position with respect to the electron beam, while the chip marks are used to locate individual chips to start the electron beam exposure of a specific pattern. As mentioned above, the EBL is often utilized for mask making for research purposes.

An innovative example of an interdisciplinary nanoscale device fabricated using the EBL is the nanoscale water transistor [8–10]. As can be seen in Fig. 3.7, both micro- and nanoscale active parts are included in this transistor. They are fabricated using the low-resolution (high-speed) and high-resolution (low-speed modes). The gap between the nanoelectrodes shown in Fig. 3.7C is around 20 nm. A smaller gap between the two

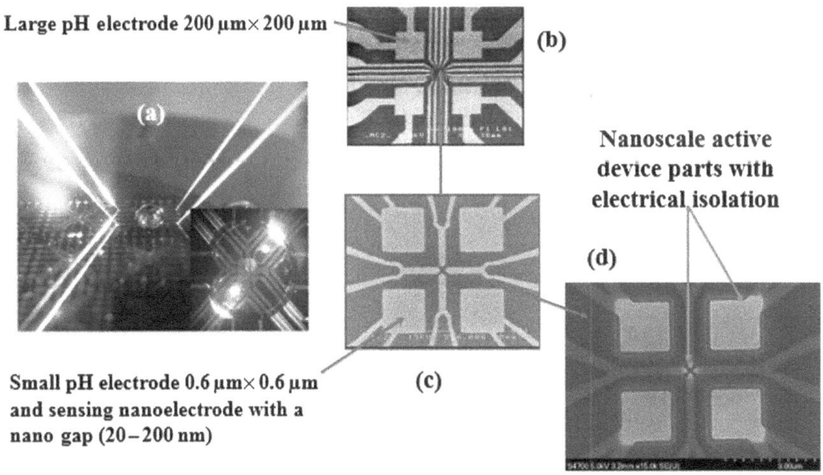

Figure 3.7 A nanoscale water transistor fabricated using electron beam lithography. (A) Photograph of the nanoscale transistor under measurements. (B) SEM image showing microscale pH electrodes. (C) SEM nanoscale gap pH electrodes located in the gap between the microscale pH electrodes as indicated. (D) SEM image of the nanoscale pH electrode with top insulation allowing contact between the water drop and the metallic electrode at a specific small nanosized area. *SEM*, Scanning electron microscope.

nanoelectrodes has also been demonstrated using EBL. As can be seen in Fig. 3.7D, with the middle area of the small pH electrodes, there is an open area where a water drop is in direct contact with the metallic electrodes. By applying a voltage (emitter to base bias), the water will decompose, and consequently the pH value will deviate from the original starting value. This will consequently lead to a modification of the emitter–collector current and hence a transistor action is observed [8]. The same prototype transistor has been used to create microvortices in water for varying pH conditions. This is of high potential and interest for biological experiments as it can be used to mimic living cell conditions [10]. Further by using sharper electrodes of the same type as those shown in Fig. 3.7C and D, and by applying an alternative bias, it was possible to use the two nanoelectrode configuration to trap single protein molecules [9]. Such prototype transistors can only be achieved by EBL top-down technique. This indicates the potential which can be achieved from such techniques. Nevertheless, the EBL being a serial process makes such devices quite expensive and does not have an economic justification for mass production.

3.3.4 Soft lithography

Soft lithography is the collective name of many newly developed techniques for the fabrication of micro- and nanostructures for many applications and specially in biological sciences. Soft lithography was suggested in 1998 [11], and it is achieved by ordinary hand work with no need for sophisticated equipment as in the case of conventional or EBL. Further, no sophisticated clean room is required for processing these small size structures when utilizing soft lithography.

All soft lithography techniques are based on printing, molding, and embossing with an elastomeric stamp [11,12]. The soft lithograph techniques allow the fabrication of small size structures in a much easier way compared to other lithography techniques. In addition, the soft lithography is low cost and as mentioned above it is suitable for biological applications. Although the technique was initially developed for structures with feature sizes in the range of 500 nm or larger, nowadays the technique has been improved to fabricate nanostructures with dimensions down below 100 nm, and is termed nanocontact printing. In soft lithography the resolution limit is determined by different actors, optical diffraction is not among them. These factors are the van der Waals forces, the wetting, and kinetic factors, that is, filling capillaries on the surface of the master. The molding material used is usually soft polymeric material. The most commonly used material for soft lithography is poly(dimethylsiloxane) (PDMS), since it is not toxic and is suitable for experiments with biological species. In this fabrication method, a precursor of the PDMS is poured over a lithographically made master, then it is cured to make it cross-linked with the sample. Finally it is peeled off. Fig. 3.8A is a schematic illustration of steps in a typical soft lithography processing. for fabricating micro- or nanosized structures using soft lithography, a master (mask or wafer) with a specific pattern is first fabricated using standard conventional (in case of microstructures) or EBL (in case of nanosized structures). It is important to note that soft lithography can produce small size structures using a variety of materials including complex molecules for biological investigations. Fig. 3.8B shows some micro- and nanosized structures fabricated using soft lithography. Although the soft lithography can even deal with curved surfaces, it is not the most suitable for the fabrication of nanoelectronics devices [13]. This is because ICs are achieved through a series of stacked layers of different materials aligned on top of

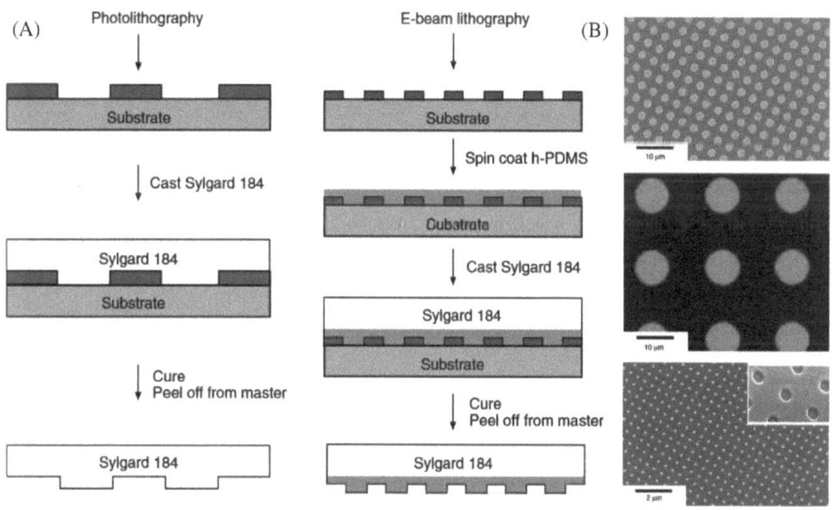

Figure 3.8 (A) Schematic illustration of the procedure for the fabrication of normal PDMS (left panel), and h-PDMS (right panel) stamps, respectively. Here the additional PDMS layer allows the fabrication of lateral feature sizes below 500 nm. (B) Typical example of micro- and nanostructures achieved using soft lithography. The top shows a scanning electron microscope of a silver microsized disks, while in the middle is a fluorescence microscope image of a microsized IgG, and bottom is a scanning electron microscope image of gold disks with a size of 100 nm. *PDMS*, Poly (dimethylsiloxane).

each other in specific locations. Hence, any deformation in the PDMS stamp can cause the misalignment of the sequences layers and hence the loss of required stacked layers will be the result. This implies that soft lithography is not ideal for configurations with different stacked layers [13]. Although researchers have replaced the soft stamp by a rigid one, still the technique is not the most ideal to use for multistacked layer configurations.

3.3.5 Nanosphere lithography

Nanosphere lithography (NSL) is a newly emerging inexpensive promising technique suitable for fabricating homogenous and regular arrays of NPs. The technique combines the advantages of both top-down and bottom-up approaches [13].

Fig. 3.9 illustrates the steps used in NSL. In principle the four steps shown in Fig. 3.9 are divided into two main parts. In (A) and (B) are the mask part and in (C) and (D) the nanostructure fabrication part. In a NSL process, the substrate surface is first coated with a suspension of nanosized spheres, usually polystyrene spheres are used, then a chemical treatment to increase the hydrophilic nature is conducted. After drying, a hexagonal closed pack (hcp) mono- or bilayer structure is then formed. This layer is called colloidal crystal mask [14]. Then through deposition, the material will be inserted into the interstices of the ordered beads. By using sonication in a suitable solvent or by stripping the nanospheres, an array of homogenous and ordered NPs will be left on the surface of the sample. Finally an annealing step is sometimes applied to crystallize the NPs or to introduce a specific required phase of the fabricated nanomaterial. The first experiments with NSL were conducted in the early 1980s [15]. With the aim of replicating a submicrometer monolayer, and by using visible light, the first NSL ordered nanostructure was introduced. The technique is then termed "naturally assembled polystyrene latex." Just a year later the scope of the technique used in [15] was extended to utilize the deposited spheres to act as a "mask" for lithography [16,17]. The simple spin-coating process is usually used for the deposition of the spheres. This techniques then became popular and it was extended to double layers of colloidal spheres and by the 1990s the technique was renamed to "NSL" [18]. Due to the high cost of conventional lithography techniques (optical and EBL, etc.) as standard techniques for patterning, the availability of this equipment in research laboratories is very limited. This has allowed NSL

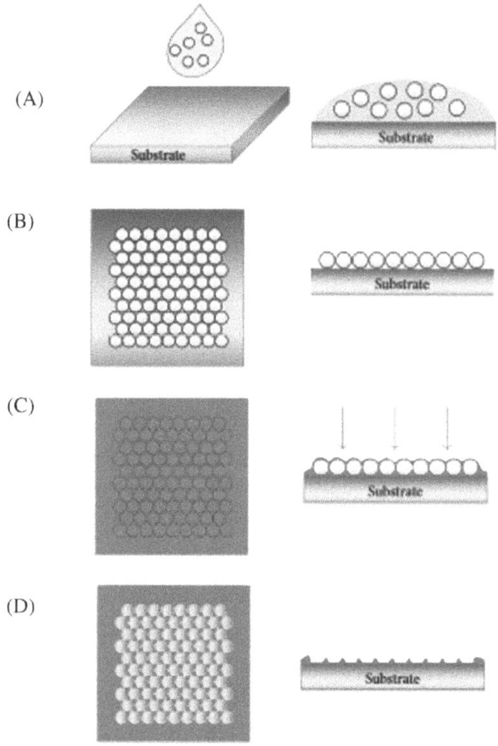

Figure 3.9 Illustration of nanosphere lithography process. In the first step (A) the deposition of colloidal particles take place, (B) the two-dimensional self-organization of the nanosphere, (C) deposition of the desired material, and (D) finally removal of the nanospheres (lift-off).

to become a popular hybrid approach (between top-down and bottom-up approaches) for many applications including [13]:

1. Those requiring nanoobjects with similar features, like size, shape, etc. (the NPs are identical and are placed at an equidistance from each other). Hence having the same properties.
2. Applications that require long-order arrangement of similar small size objects.
3. Applications that require control of the lateral density of the deposited nanosize particles.

In addition, the NSL offers flexibility of the material type involved for both the substrate and the deposited material. In addition, it is a fast and relatively cheap approach [13].

3.3.6 Colloidal lithography

Colloidal lithography is based on the use of colloidal crystals for masking, etching, deposition, and it facilitates the fabrication of different nanostructures on different types of surfaces [19]. The colloidal lithography shares the same interesting characteristics of being high throughput, and a diverse choice of materials can be processed. All these interesting advantages are achieved with low cost. In this connection the NSL is a special case of colloidal lithography, where colloidal spheres specifically are used for the masking material.

In a colloidal mask experiment usually evaporation, sputtering deposition, chemical deposition, chemical vapor deposition (CVD), etc., are the typical methods used to deposit an array of nanostructured material. Both inorganic and organic arrays of nanostructured materials have been demonstrated using the colloidal lithography. Even an organic light-emitting diode (LED) structure has been demonstrated using colloidal lithography [20]. The development of such organic LEDs adopting such simple and cheap method, in fact provides a shortcut to fabricate optoelectronic devices [20]. Due to the simplicity, low cost, and the possibility of depositing many different three-dimensional (3D) arrays of nanostructures of a variety of materials using the colloidal lithography, many processing routes have been developed, for both the masking and the deposition of the material. Vertical deposition, dip-coating, spin-coating, nanorobotics manipulation, and template-assisted epitaxial growth have been demonstrated for mask making processes [21–25].

Further, by varying the configuration of the colloidal NP mask, different morphologies of the nanosized particles can be achieved. As can be seen in Fig. 3.10, by using single or double hcp NPs-based mask, shown in part (A), a single layer periodic triangular nanosized array can be fabricated as seen in part (B) of the figure. The coverage of the nanosized triangular in this case is about 7.5% of the total surface area of the substrate [26]. While when using double layer colloidal hcp nanosphere as the mask configuration, then the fabricated nanosized array will consist of circular NPs [26].

Even for depositing the selected material for obtaining an ordered nanosized array of particles, different approaches have been adopted. When the colloidal lithography techniques were introduced, the nanosized particles that were fabricated comprised nanotriangles similar to those shown in Fig. 3.10B.

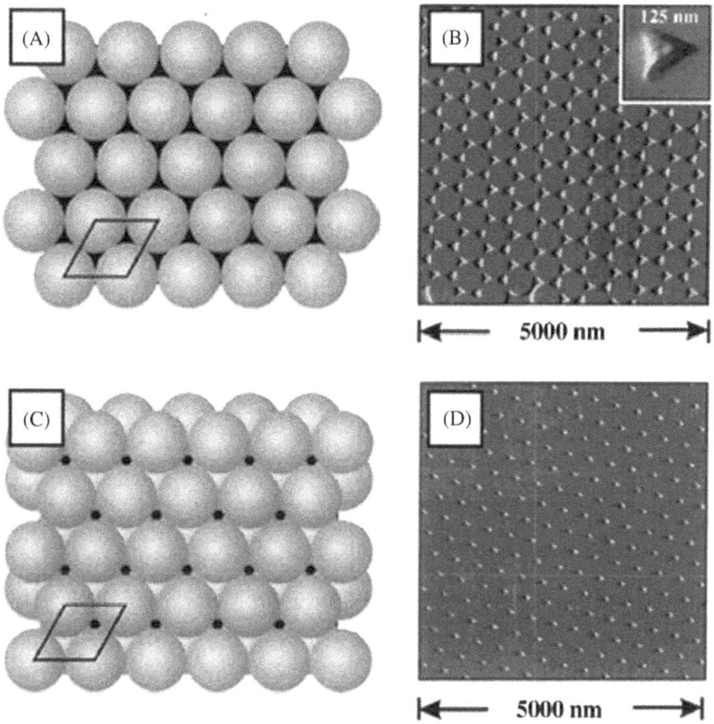

Figure 3.10 (A) Atomic force microscope micrograph of single layer of colloidal hcp mask. (B) The corresponding nanosized particles obtained using the mask shown in (A). (C) Atomic force microscope micrograph of a double layer of hcp colloidal mask. (D) The corresponding nanosized particles fabricated using the mask in (C). *hcp*, Hexagonal closed pack.

By resolving the angle of the incident material with respect to the mask orientation in colloidal lithography (ARCL), and applying multidepositions, different interesting morphologies can be fabricated [26]. By applying a nonzero incident angle with respect to the normal to the substrate (see Fig. 3.11A), it was possible to reduce the size of the deposited NPs by a factor of four. Further, using the ARCL, deformed shapes compared to the morphology obtained by zero angle incident can easily by achieved. In addition, new nanostructures such as nanooverlaps, nanogaps, and nanochains have all been demonstrated using ARCL [26]. It is also to be noted that the different processes of deposition of different materials is independent of each other. This implies that ARCL can be used for deposition, it is possible to fabricate a variety of binary of materials. This is a unique advantage of the colloidal lithography as such binary

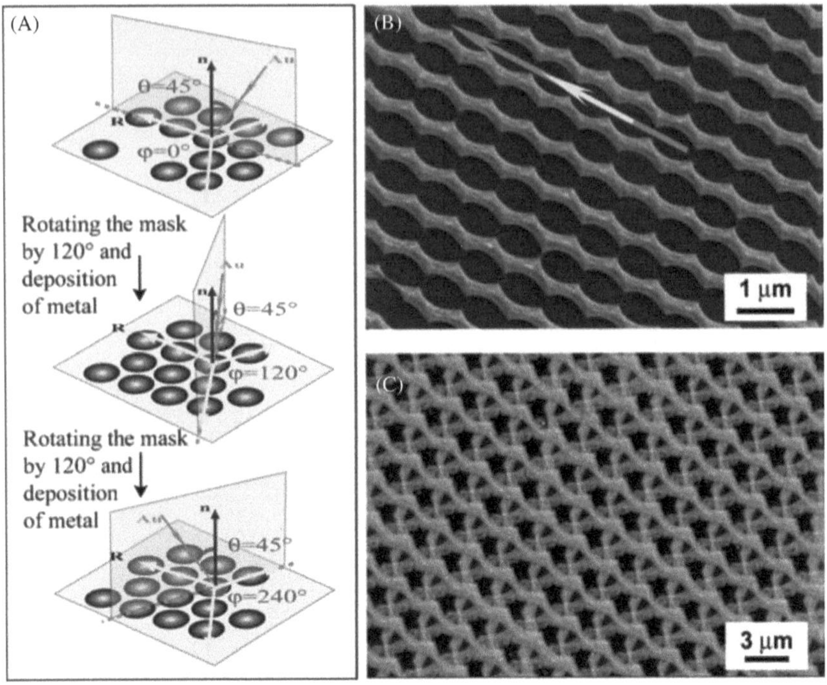

Figure 3.11 (A) Schematic depiction of the angle resolved colloidal lithography. (B and C) The nanosized structures obtained by one and three depositions in colloidal lithography processes, respectively.

combinations of different materials are not easy to fabricate using standard lithography [26]. The depiction of the ARCL is shown in Fig. 3.11A. By varying the angle of incident of the deposition beam, different 2D and 3D arrays of complicated nanostructures grids can be easily fabricated. As can be clearly seen in Fig. 3.11B, zigzag nanowires 2D arrays have been fabricated. Further, by modifying the angle further, it is possible to fabricate a 3D complex grid of nanowires of different materials stacked on top of each other as shown in Fig. 3.11C. The independence of the chemical properties of different deposited materials in different ARCL steps allows the design and fabrication of a variety of different materials of interest. Such a possibility is not straightforward using conventional or other soft lithography techniques [19]. Due to the recent development in the processing using colloidal lithography, namely, introducing etching, incident angle resolved colloidal lithography, stepwise and regular mask registry modifications have all made the colloidal lithography to be a powerful tool for nanochemical patterning [19]. Hence, the colloidal lithography technique provides some unique advantages and disadvantages. The advantages are low-cost budget operation, high throughput, and easy processing with no need for advanced instrumentation. It is also suitable for planar and curved surfaces. By reducing the colloidal particle (spheres) sizes, NPs and cross-linked structures with sizes below 100 nm are easily fabricated. Nevertheless, the fabrication of spheres with dimensions less than 100 nm is a challenging task. In turn, this limits the minimum feature size of the deposited nanosized array of particles to be larger than 10 nm. Despite the many positive aspects of colloidal lithography, there are still some obstacles to widen the application scope of colloidal lithography. Among these obstacles is the fact that creating defect-free single crystals by using colloidal lithography is not easily achieved. To process the deposited nanosized particle for crystallization and transfer it to defect-free structures will in most cases cause damage the pattern's regularity. Another critical disadvantage is the difficulty in the integration of the colloidal lithography with conventional standard lithograph techniques for microelectronics and nanoelectronics [19,26]. These disadvantages have limited the use of colloidal lithography being of interest for fundamental studies of chemists, biologists, and physics but not for creating electronic or optoelectronic devices, which are the main driving factors for the development of nanotechnology.

3.3.7 Scanning probe lithography

Although scanning probe (tunneling) microscopy (STM) and atomic force microscopy (AFM) emerged as atomic resolution imaging techniques, they have been used to fabricate small size nanostructures. The STM is a technique based on the use of a small size tip to image the surface at atomic resolution. Usually the tip has a size less than 50 nm. Due to this small size tip, the STM and AFM have been suggested as fabrication tools for small size structures due to their ability to manipulate material at nanometer sizes. The use of STM and AFM for manipulating and fabricating nanosize structures is termed scanning probe lithography (SPL) in the case of using STM, while when using ATM to selectively remove material, it is called dip-pen nanolithography and sometimes this technique is termed dip-pen nanodisplacement lithography (DNL). The STM and AFM constitute high-resolution direct writing techniques with the ability to generate arbitrary patterns as will be shown below. These are advantages, but the disadvantages are the fact that they are both serial techniques with relatively low processing speed.

When a surface is approached by an STM tip, a finite force is exerted onto the adsorbed atom on the surface. This force is a combination of both Van de Waals and electrostatic contributions [27]. This force can be tuned in magnitude and direction, this is achieved by applying the proper tip voltage at a specific chosen location. It is obvious that lower force is required to move the atom compared to the force require to completely pull and remove the atom from the surface. Hence, by an optimized STM tip applied force, atoms adsorbed to the surface can be manipulated and moved around on the surface. In 1990 at IBM laboratories, STM tip was used to manipulate and move atoms adsorbed to the surface of a substrate and to perform direct writing on the surface using single atoms [27]. These experiments were performed under ultrahigh vacuum (UHV) at a relatively low temperature of around 4K. The UHV and low temperature are necessary for such experiments to prevent any contamination on the surface. Without such conditions, it is impossible to perform clean clear direct writing using atoms to build letters or any arbitrary geometry. In addition, another important requirement for writing using atoms by an STM tip is that the corrugation of the surface potential is to not exceed or be below certain values [27]. Xenon (Xe) atoms placed on the surface of a nickel substrate will satisfy the required corrugated surface potential. These experiments were successful in demonstrating direct writing using atoms [27]. Using an STM tip to write using single atoms will imply that each letter will have a size of around 1 nm. This will imply that almost all the books written in the world can be rewritten using an STM tip and proper atoms on a single A4 page. Nevertheless, this process will probably take an infinite time and the atoms text can only be read using STM. This renders the technique to be unuseful if such obstacles are not removed.

On the other hand, dip-pen nanodisplacement has emerged as a promising desktop technique for nanofabrication [28].

To enable the DNL as a mass production low-cost technique for the fabrication of nanostructures, it was suggested to increase the tips used in the process. A recent report demonstrated the use of 55,000 tips for the fabrication of gold dot arrays [28]. This implies high throughput and can lead to the mass production of NPs at low cost. The difference between the DNL and the SPL is that in the former case the process is performed at ambient conditions, while for the latter an UHV and low temperature are necessary for achieving the direct writing. Fig. 3.12 shows the different steps for producing gold NPs. Again, and as a disadvantage, it is not yet

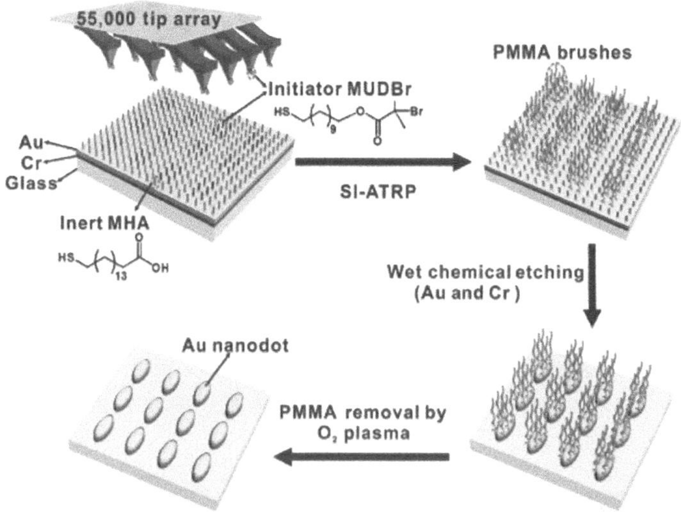

Figure 3.12 Illustration of multitip DNL process steps for the fabrication of nanostructures. *DNL*, Dip-pen nanodisplacement lithography.

clear how this technique can be integrated with other standard lithography techniques used in the electronic industry. This in turn limits the scope of the possible applications for nanostructures fabricated using the DNL.

3.4 Bottom-up conventional nanofabrication methods

By bottom-up nanofabrication, we mean methods that rely on the fabrication of nanostructures by assembling small size structures starting from single atoms. By conventional we here target those methods performed at elevated temperatures, that is, at temperatures of a few hundred degrees Celsius. Further, this category is divided into either physical or chemical-based processes. The physical methods are usually performed in gas phase while chemical methods are performed in liquid phase. We will restrict the presentation in this chapter to physical (gas phase) methods. The chemical methods (usually performed at much lower temperature) will be presented and discussed in the coming chapters as they are the main topic of the this book.

3.4.1 Vapor liquid solid synthesis of nanostructures

The vapor liquid solid (VLS) synthesis was developed more than 50 years ago. In 1964 Wagner and Ellis [29] suggested the VLS technique and they

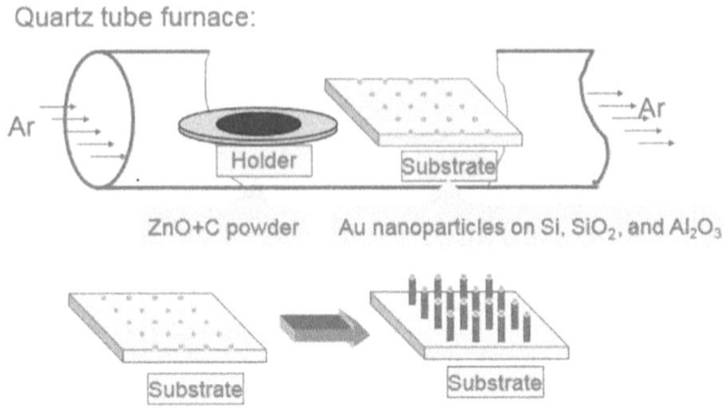

Figure 3.13 Schematic representation of the vapor liquid solid for the synthesis of nanowires.

demonstrated the synthesis of silicon microwhiskers. However, as of today the technique is used routinely to synthesize different nanostructures. Although the VLS technique is simple with no need for complex instruments, it requires relatively high temperatures in access of 800°C [29,30]. Fig. 3.13 shows a schematic diagram of a VLS system. To fabricate nanostructures by VLS, a clean substrate is first coated with a thin metallic catalytic layer. Usually a thin gold (few nm) layer is first deposited on the substrate.

Then a powder of a specific material is placed in a clean quartz tube. Beside this material, and after it, a substrate with the thin catalytic metallic layer is placed. Argon is used as the carrier gas. Depending on the required material to be synthesized sometimes an additional gas, for example, oxygen or any other gas, is also introduced beside the argon. The temperature of the synthesis is usually higher than 800°C. Due to this high-temperature requirement, the VLS technique is very sensitive to contamination inside the quartz chamber. Nevertheless, as today, the availability of clean-room laboratories with controlled environment, the possibility to obtain high-quality quartz furnace tubes, and the high-grade chemical powders have enabled the VLS technique to be suitable for the production of pure and high-quality nanostructures of different morphologies. In general, synthesis starting from the gas phase at high temperatures is usually preferred due to the expected high crystal quality. In most of the VLS processes, the growth will lead to nanowires/nanorods due to the fact that vertical growth is energetically favorable compared to lateral growth. A disadvantage of the VLS is the restriction of the substrate due to the high temperature. Only materials that

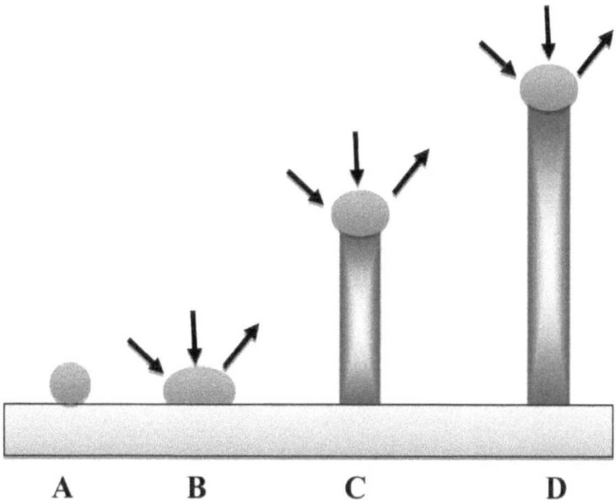

Figure 3.14 Illustration of the growth mechanism of nanowires using the vapor liquid solid technique.

withstand such relatively high temperatures can be used. The mechanism of the growth for the VLS technique is schematically shown in Fig. 3.14. As the temperature inside the quartz tube rises, the catalytic metal starts to melt and forms a droplet (Fig. 3.14A). At the same time, the powered material starts to evaporate and is transformed from the solid to the gas phase. The vaporized material will then be carried by the argon gas. Due to thermodynamics considerations, the vaporized material will then be adsorbed at the liquid surface (catalytic metal droplet), and will diffuses into the droplet (Fig. 3.14B). This is then followed by supersaturation and nucleation at the liquid/solid interface leading to axial crystal growth. The axial growth dominates over lateral growth due to energy considerations (Fig. 3.15).

The VLS method has been used for the synthesis of 1D nanostructures, like nanowires, nanorods, nanotubes etc., for a variety of materials. Among these materials is zinc oxide [29,30]. To synthesis ZnO nanowires, the synthesis is achieved by using ZnO powder. Since the melting point of ZnO is above 1900°C, ZnO powder is mixed with graphite to reduce the melting point to around 800°C. Fig. 3.1 shows some scanning electron microscope (SEM) and transmission electron microscope images of ZnO nanowires grown by the VLS technique [29]. As can be seen ZnO nanowire grows only on the places with the Au catalyst. The grown ZnO nanowires in this study have a diameter ranging between 20 and 150 nm and lengths of around 10 μm. They are oriented along the 0001 direction,

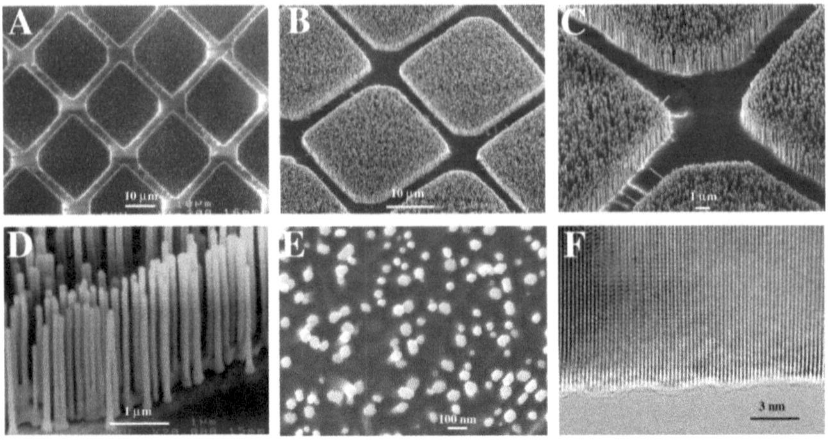

Figure 3.15 (A–C) Top view SEM of the well-faceted hexagonal ZnO nanowires. (D) Side view SEM of the synthesized ZnO nanowires. (E) SEM of the Au 10-35 Å layer after annealing. (F) High-resolution TEM image of an individual ZnO nanowire showing its 0001 growth direction. For the nanowire growth, clean (110) sapphire substrates were coated with a 10–35 Å thick layer of Au, with or without using TEM grids as shadow masks (microcontact printing of thiols on Au followed by selective etching has also been used to create the Au pattern) [29]. *SEM,* Scanning electron microscope.

that is, vertically oriented. A surface emitting laser at 385 nm was demonstrated using these ZnO nanowires [29]. The diameter and length of the produced 1D nanostructures can be varied using different synthesis parameters, like catalytic layer thickness, growth duration, temperature, etc. Although single crystal nanostructures of different materials were demonstrated using the VLS technique, the relatively high processing temperature and the existence of a metal on top of the nanostructure limit the use of substrates to those withstanding such high temperature. Also at such high temperature, the problem of contamination must be carefully considered. In a special case of the VLS technique, one of the constituent atoms of the material to be grown can act as a catalyst. This is called self-catalytic growth [31]. In this case the foreign catalytic material usually needed in the VLS growth will be eliminated. Nevertheless, the catalyst will remain on the top of the nanowire as in the case of the conventional VLS growth process. Further, another version of the VLS technique, named vapor solid (VS) growth has also been developed. In the VS growth process, a spontaneous condensation of the vapor into solid material occurs due to the decrease of the Gibbs free energy or the reduction of the supersaturation state of the vapor [32].

3.4.2 Chemical vapor deposition for the fabrication of nanostructures

In general thermal evaporation is one of the simplest single crystal growth methods. Although the technique is old and has been developed for the growth of many semiconductor material thin films, it has been used successfully for the synthesis of different 1D nanostructures, like nanorods, nanotubes, etc. The basic process in CVD is based on the sublimation of a source material at a high temperature, usually at low pressure. The source material is either in solid or powder forms. The sublimation is then followed by deposition on a substrate forming the required material. The CVD technique is commonly used in the industry as it can be designed for mass production. A typical CVD system is composed of a chamber made of quartz or alumina. This chamber is connected to pumps to control the pressure inside the chamber. The chamber can be horizontal or vertical, and the source material is placed in the middle of the chamber while the substrate is placed at the downstream end of the chamber. The growth process starts by pumping down the chamber to around 10−2 Torr. Then the heating is turned on and the temperature rises to a specific reaction value. Usually an inert gas, like argon or nitrogen is then introduced in the chamber—it is designated as a carrier gas. The temperature and pressure are then kept constant for a certain period of time. The vaporized source material(s) are then carried by the carrier gas to the low-temperature region where it becomes supersaturated and will be deposited on the substrate forming the required material. There are many different types of CVD chambers with processing at different gas pressures. In most of the depositions using the CVD, a flat substrate is used. After the exposure of the surface of the substrate to the chemical vapor, an initial monolayer of atoms or molecules is deposited, and will act as a nucleation layer for subsequent growth as long as the vapor of the reactants is at supersaturation and the temperature is suitable for the specific material growth. Although CVD was initially developed for the growth of relatively thin layers of semiconducting materials for industrial mass production in the microelectronics industry, it is now emerging as a method for the synthesis and production of 1D nanostructures like nanowires and nanotubes and even for 2D nanostructure materials.

Nanostructures of different materials have been synthesized using CVD in recent years, for example, ZnO nanowires, boron nitride nanotubes, carbon nanostructures [33−35]. Although most of the demonstrated nanostructures synthesis using CVD were achieved on template-free substrate, a few investigation on the synthesis of nanostructures on substrates with a preprepared template have indicated the visibility of patterned substrates for the synthesis of nanostructures on a specific locations, that is, selective area epitaxy (SAE) [36]. An example of the synthesis of p-type doped ZnO nanowires array is briefly presented and discussed below [33]. As shown in Fig. 3.16 above highly dense vertically aligned thin (around 55 nm) ZnO nanowires have been synthesized by a

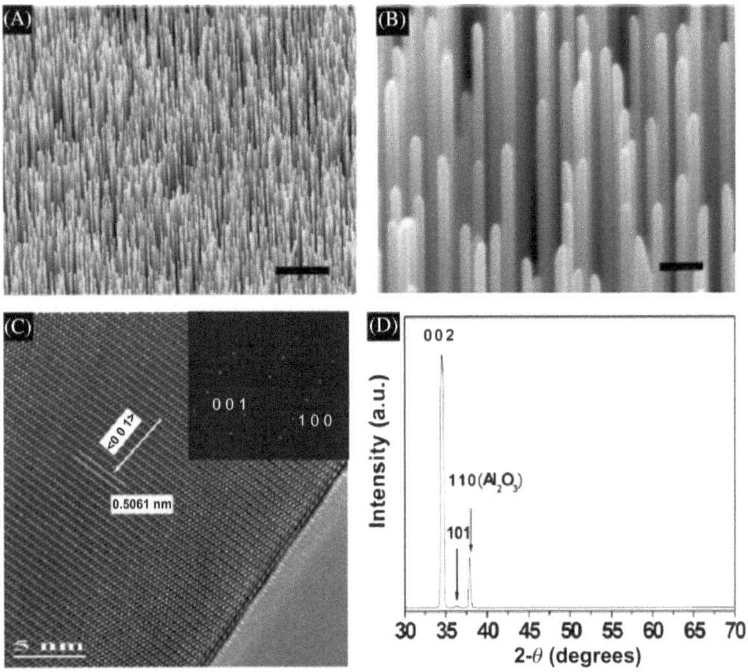

Figure 3.16 Structural and morphological characterization of the ZnO:P NWs. (A) Low-magnification SEM image of the well-oriented ZnO:P NW arrays. Scale bar is 2 μm. (B) High-magnification SEM image of the ZnO:P NW arrays, showing that the NWs have uniform diameters of about 55 nm and very smooth surfaces. Scale bar is 200 nm. (C) HRTEM image of ZnO:P NWs showing high-quality single crystal with extremely clean and smooth surface; note the absence of any amorphous layer coating on the surface. The electron diffraction pattern is shown in the inset, from which no second phase or cluster could be detected. The growth direction is along 001 direction as indexed in the images. (D) XRD spectrum of the ZnO:P NWs treated by RTA. The typical X-ray diffraction pattern of ZnO NW arrays is observed, with no peaks associated to second phases or clusters. *HRTEM*, High-resolution transmission electron microscope; *NW*, nanowires; *RTA*, Rapid Thermal annealing; *SEM*, scanning electron microscope; *XRD*, X-ray diffraction.

simple CVD method. Further these ZnO nanowires were p-type doped. In a growth process using a simple tube furnace CVD process with a mixture of O_2 and N_2 used as carrier gas, both unintentional and p-type doped ZnO nanowires of high quality were synthesized [35]. A mixture of pure (99.999%) ZnO and Zn-carbon (1:1) powders were mixed together and were used as the growth source material for the synthesis of ZnO nanowires. For p-type doping P_2O_5 powder was used together with the source material for the case of p-type doped ZnO nanowires. The source and doping powders were put into an Al_2O_3 boat that was placed in the middle of the tube, and a sapphire substrate was put at a distance of 4.5 in. in the downstream direction of the furnace. Then the temperature was raised to 945°C and the growth duration was 30 minutes. For the growth of the p-type ZnO nanowires, ZnO/graphite was mixed to a 1:1 ratio using N_2 gas only as the carrier gas. Fig. 3.16 shows different SEM, TEM, and X-ray diffraction results indicating that high crystal quality has been achieved [35].

Bearing in mind that CVD systems are expensive, and the fact that the synthesis of nanostructures is usually performed at temperatures larger than 900°C, which implies that extra care has to be taken to avoid contamination, the use of the CVD technique has been limited. In addition, it is important to mention that during deposition, material is deposited everywhere in the chamber, that is, the walls of the CVD chamber will be contaminated. Due to this contamination, the chamber/furnace walls have to be cleaned routinely. Due to all these factors, today the number of papers published on the synthesis of nanostructures using CVD has been reducing.

3.4.3 Molecular beam epitaxy for the fabrication of nanostructures

Molecular beam epitaxy (MBE) is a complex UHV growth system, which is capable of depositing atomic layers at a time and can produce high-quality crystal films. It operates at rather relatively high temperatures, but usually lower than that for CVD systems. It is a sophisticated thermal evaporation system where different molecular beams, that is, more than one beam, are emitted from a heated source and target toward a heated substrate. MBE system development started in the late 1960s [37]. A typical MBE system is shown in Fig. 3.17.

Since atomic layers are deposited at a time, the MBE is a growth system with high control over the thickness, composition, and doping of the

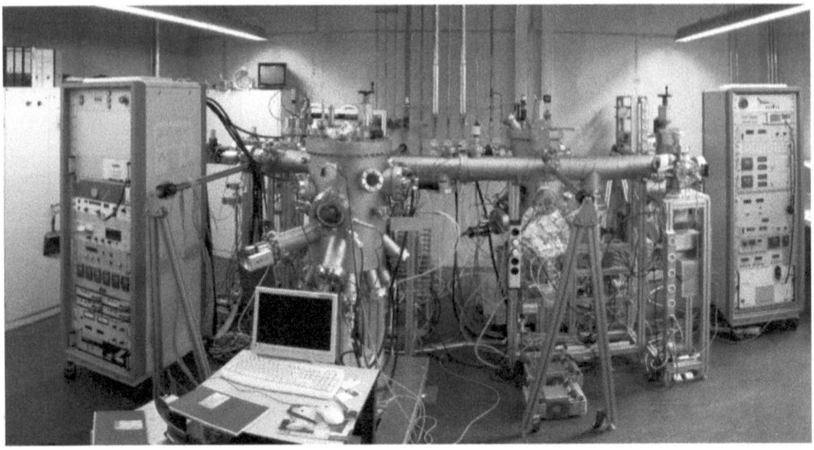

Figure 3.17 Typical molecular beam epitaxy system installed at Goethe University Germany.

grown films. Usually some surface analysis characterization tools are connected to the MBE chamber in order to monitor the quality of the grown films. MBE has been very successful in demonstrating abrupt interfaces between different lattice mismatch semiconductors [38]. In addition, and due to the high degree of control in MBE systems, the technique has been successfully used in growing nanostructures, especially 1D morphologies. Although nanostructures in general and 1D specially do not suffer require lattice matched substrate, mainly due to their small footprint, MBE have been successful in demonstrating abrupt interface between different nanowires and lattice mismatched substrates [38]. In general the use of MBE as a technique to grow nanostructures requires substrate special treatment. In principle there are three possible routes that are being popular for the growth of nanostructures using the MBE technique. The first is the use of patterned substrates, that is, SAE. Here a passive layer of SiO_x or SiN_x etc. is first deposited and patterned using EBL to create nanosized holes of a specific required geometry [39]. The second approach is the use of nanowires arrays as templates of subsequent growth. In this case a substrate with already grown nanowires or nanopillars is used to overgrow another material. An example of this is the growth of AlGaN nanowires on GaN nanopillars [40]. Since a nanopillar template substrate will lead to a shadow effect on adjacent nanowires, synthesis of overgrowth of another material's nanowires is achievable just by using a substrate with a template [40]. Semiconducting heterostructures research on III–V and II–VI materials have been an active research area from the 1980s. Nevertheless, the unacceptable density of defects and dislocations, usually higher than 10^8 cm^{-3} [41], has hindered the progress of developing mature devices. Despite the unique properties that can be gained from these heterostructures, the absence of suitable substrates has been the main reason for the presence of unacceptably high levels of defects and dislocations. This is due to the strain/stress at the film/substrate interface. However, nanostructures, for example, nanodots and nanowires, provide a possible solution to the reduction of these defects and dislocations that deteriorate performance, since they have a relatively small "footprint" and can relax the strain/stress at the substrate interface. This fact has encourage researchers to try to integrate and use nanostructures, and especially 1D morphologies, for different device applications [41]. Hence many different laboratories have initiated research to achieve and develop practical growth processes for such semiconductor heteronanostructures for different device applications. The focus was concentrated on 1D nanowires as they can be efficient in relaxing the strain at the interface with the substrate [41]. Nevertheless, and although nearly defect-free AlGaN and InGaN suitable for UV LEDs have been demonstrated, these efforts have been again hindered by the lack of a suitable transparent soft material for planarization. In addition, there is large variation in the thickness [42–44]. Other researchers have relied on the growth of nanowires by coalescence to form a suspension that can be suitable as virtual substrate/template for the epitaxial growth for different photonic and electronic devices with no need for planarization [40,45,46]. Nevertheless, such an approach has proven to lead to no control over the size, morphology, and orientation of the grown nanowires and hence no control over the observed properties [41].

Instead, the research efforts have concentrated on the utilization of patterned substrates to achieve controlled SAE [41]. As mentioned above there are two different routes for such processes. These are shown schematically in Fig. 3.18A and C. Using EBL, first a masked pattern is prepared. Then this EBL patterned mask is used for the SAE of GaN high oriented GaN nanowires. In this case, the nanowires morphology is highly controllable as shown by the SEM in Fig. 3.18B.

Although the MBE system provides high control, since it can be used to grow a few monolayers on a substrate, the technique cannot be used to directly grow nanostructures. For nanostructure growth with MBE, a patterned layer that enables selective epitaxy is essential. For this patterning,

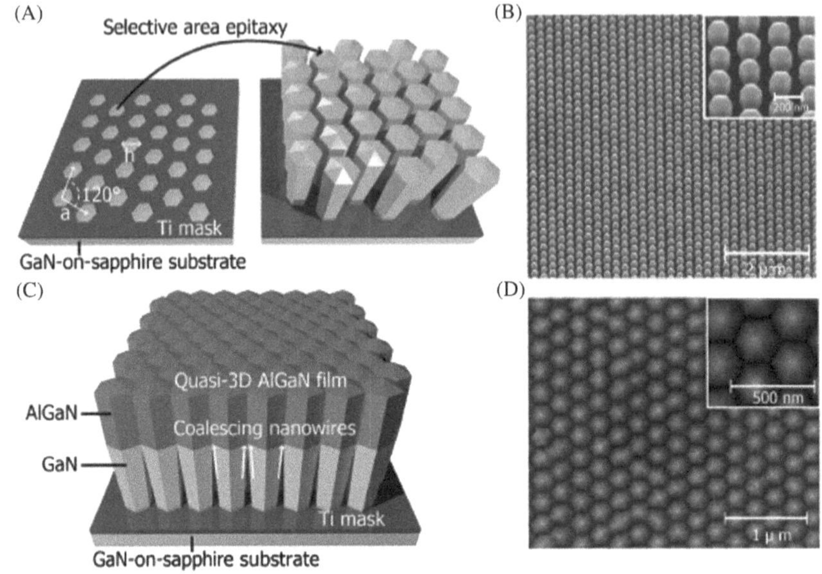

Figure 3.18 (A) Left: Arrays of nanoscale opening apertures defined by e-beam lithography process on a Ti mask on planar GaN template on *c*-plane sapphire substrate. Right: Schematic of the selective area epitaxy of GaN nanowire arrays on the nanopatterned substrate. (B) 45-degree-tilted view SEM image showing GaN nanowire arrays selectively grown in the opening apertures. Inset: High-magnification SEM image of GaN nanowire arrays. (C) Illustration of the concept of gradual coalescence process that leads to the formation of semipolar AlGaN film structures. (D) Top view SEM image of the AlGaN film structure formed through controlled coalescence of AlGaN nanowire arrays grown on sapphire substrate with a lattice spacing $a \approx = 250$ nm and a lateral size $h \approx 180$ nm for each opening. Inset: High-magnification SEM image of coalesced AlGaN nanowire arrays [41].

holes or other geometries of nanometer sizes are needed and this implies the need for EBL. Since the EBL is a serial based technique, the cost to perform SAE will hence be relatively high. Add to this the complexity due to the UHV requirements, and the use of the MBE system is quite limited, and only a few papers have been published in previous years using this technique.

3.4.4 Pulsed laser deposition of nanostructures

Pulsed laser deposition (PLD) is another commonly used high-temperature technique developed for the growth of thin films of different materials. It is a physical vapor deposition technique in which a high-power pulsed laser is directed toward a target material with the aim of decomposing the target material to be vaporized, that is, to form a plasma plume [47]. This plasma plume is to be deposited on a nearby substrate. The PLD chamber can be operated at an UHV or in the presence of some background gas(es). PLD is frequently used for the deposition of a variety of oxides and in that case oxygen is used as the background gas. This is to ensure the complete oxidation of the deposited material. The PLD system is much less complex compared to other physical deposition systems like the CVD or MBE. Nevertheless, the process of laser ablation is quite complex. As the laser strikes the target material, its energy is transferred as electronic excitation which is then converted into thermal, chemical, and mechanical energy which leads to ablation, plasma formation, and evaporation of the constituents of the target material. The evaporated material spreads then in the chamber as an energetic plume species composed of electrons, atoms, molecules ions, clusters, and globules, that reach the substrate, which is heated to a specific temperature determined by the growth properties of the material to be deposited. As mentioned above the whole of process of PLD is complex in nature, including the laser ablation and the plasma plume including the high energetic species. In principle the PLD is dependent on five different parts that have to be optimized carefully for the growth of high-quality crystal materials. These are:

1. Absorption of the laser by the target material
2. Laser ablation of the target material and the formation of the plasma
3. Plasma dynamics
4. Deposition of the ablation material on the substrate
5. Nucleation and growth of the required film

All the above stages are critical for the formation of high-quality crystal. As these steps are complex in nature, researchers use Monte Carlo simulation to decide and optimize the operation processing parameters [48] (Fig. 3.19).

Although PLD was developed for thin-film deposition, it has been used to fabricate and grow nanostructures [49–51]. The use of PLD for the growth of nanostructures dates back to the late 2000s [49]. Nevertheless, the growth of nanostructures using the PLD is not straightforward and very few have managed to achieve it without any template layer. In some of the published experiments for the growth of ITO nanostructures using the PLD technique, it was found that it was possible to achieve nanowires using an excimer laser at a temperature of 500°C using nitrogen as the gas for the background atmosphere [52]. In this

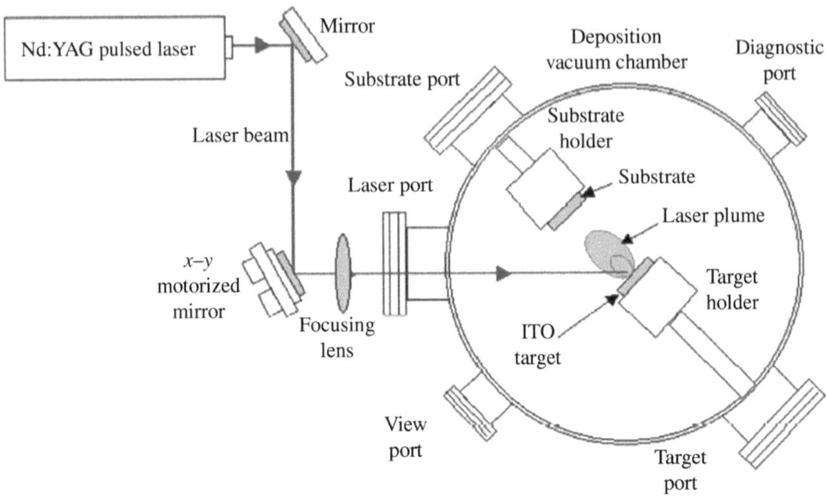

Figure 3.19 Typical pulsed laser deposition chamber for the growth of indium tin oxide.

study the authors report self-nucleation growth without the need for any pregrowth substrate preparation. Nevertheless, these results were not possible to repeat in other laboratories [49].

As shown in Fig. 3.20, it was found that at a specific narrow pressure range, it is possible to grow pure ITO nanowires. The mechanism of achieving these ITO nanowires was explained as being related to the VLS growth [49]. At low pressure, a large amount of liquid is formed. This is proposed to promote the growth of thin-branched nanowires. By increasing the pressure, the amount of the liquid formed in the chamber will be reduced. The increased pressure was found to lead to a decrease in the liquid phase inside the PLD chamber. Consequently, the number of branches were found to be less and less. At a pressure of 1 mbar, the branches around the ITO nanowires were found to disappear and it was possible to achieve pure ITO nanowires [49,52]. As the pressure is further increased to 2 mbar, it was not possible to achieve the nanowires morphology in new growth runs, and instead a thin film composed of columns with relatively large pyramids was observed. This indicates the narrow pressure range for growing pure ITO nanowires. As mentioned above it was not possible to repeat these results which report self-nucleation in a template-free growth of ITO nanowires using PLD in other laboratories.

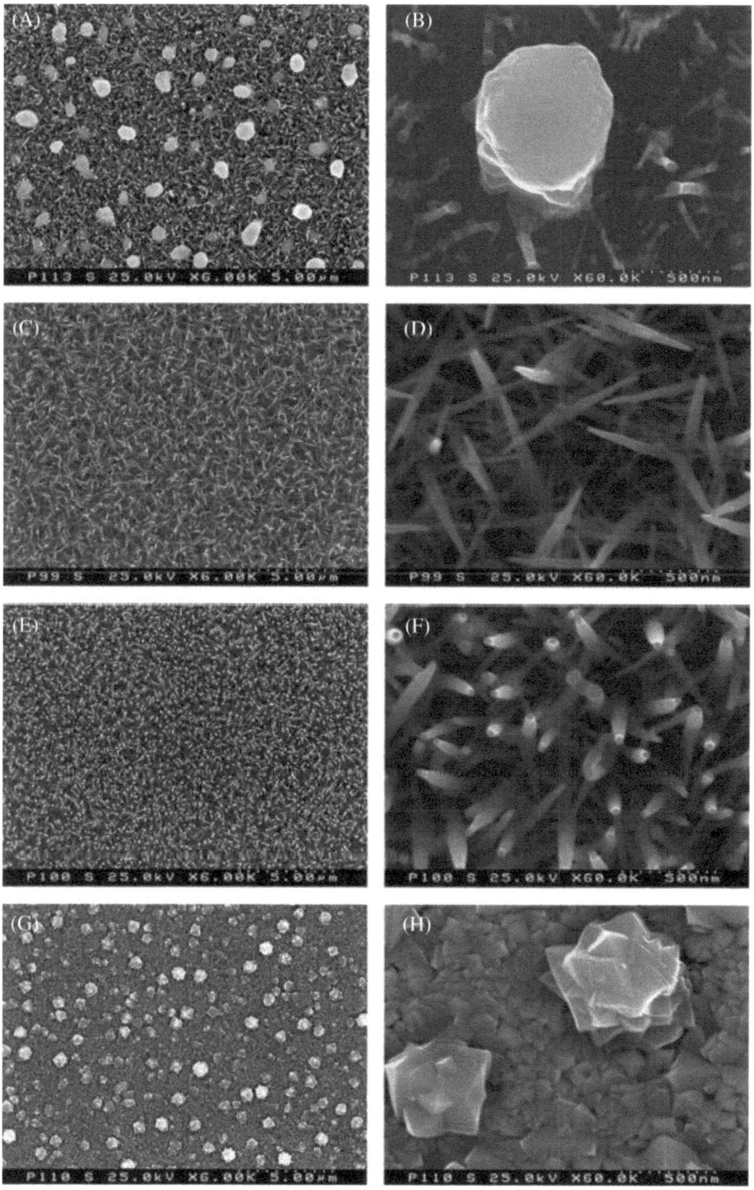

Figure 3.20 SEM surface images of nanostructured films grown using the pulsed laser deposition technique at 0.1 mbar (A, B), 0.5 mbar (C, D), 1 mbar (E, F) and 2 mbar (G, H) [52]. *SEM*, Scanning electron microscope.

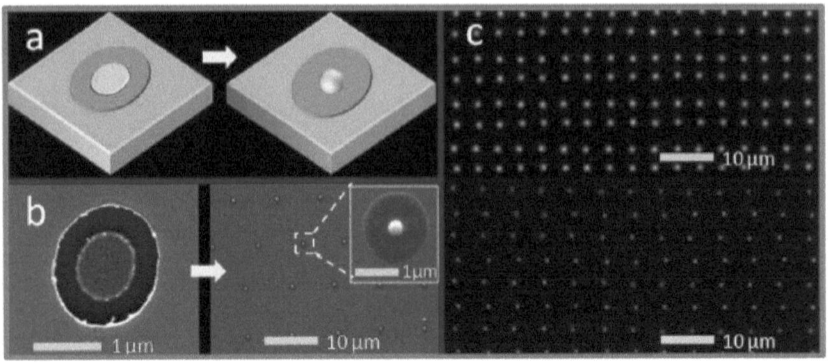

Figure 3.21 (A) Schematic diagram depicting the single nanoparticle formation under dual-wavelength femtosecond laser irradiation from an Au film on a dielectric/semiconductor substrate. (B) SEM images of the concentric structures and Au nanoparticles on Si substrate fabricated by dual-wavelength fs laser irradiation of a 30 nm Au film at pulse energies of 0.012 and 0.024 μJ. (C) Array of Au nanostructures fabricated by this method and visualized with DF microscopy; the pulse energies were 0.021 and 0.026 μJ for the upper panel and lower panel, respectively [51]. *SEM*, Scanning electron microscope.

As can be clearly seen from the previous discussion, it is not straightforward to obtain regular uniform nanostructures using standard conventional PLD on a bare substrate. Even though some researchers have published the synthesis of nanostructures using conventional PLD on bare substrates, it was not possible to repeat their results in other laboratories. Nevertheless, using a template or other intermediate nanoscale mask, some recent research has achieved nanofabrication using the PLD [50,51]. A nanostencil technique was used to pattern a substrate, and then deposition of complex oxide materials, for example, $BaTiO_3$ was demonstrated [50]. Nevertheless, postgrowth high-temperature annealing at 990ºC was necessary to obtain a polycrystalline structure. More recently, and by using complex femtosecond (fs) dual-wavelength laser, Au NPs were demonstrated [51]. The results of this investigation are shown in Fig. 3.21. The fabrication of these plasmonic Au NPs was based on the dewetting effect, and it was possible to achieve NPs by a lithography-free process [51].

3.4.5 Sputtering growth of nanostructures

Growth of nanostructures by sputtering technique is another high-temperature technique. To remind the reader, by high temperature and throughout this book, we mean processing temperature well above 100°

C, that is, a temperature of a few hundred degrees Celsius, whereas low temperature refers to growth processes performed at temperatures less than 100°C. There are quite a few recent investigations into the use of sputtering techniques for the fabrication of nanostructures. In general, sputter deposition is a physical vapor deposition approach. In sputter deposition material is ejected form a source target material toward a substrate. The ejected atoms from the source material have an energy that usually spreads in a wide range. Sputtering sources usually employ magnetrons which rely on magnetic and electrical fields to focus plasma particles near the surface of the source material to be sputtered. Under the magnetic field, electrons will move in helical paths leading to more collisions near the surface of the target material. There have been a few investigations published on the successful growth of nanostructures using the sputtering approach. Zinc oxide (ZnO) which is among the most researched metal oxides has a large share of these published papers [53–55]. Fig. 3.22 shows different SEM images of ZnO self-assembled 1D nanowires grown by rf magnetron sputtering [55]. The self-assembled 1D ZnO nanowires were grown on silicon substrate at 650°C. The deposition in this experiment was achieved under various argon sputtering pressures by the rf magnetron sputter deposition technique. The source material was pure (99.999%) ZnO powder. Before deposition, the substrate was subjected to standard silicon cleaning steps. The sputtering chamber was first pumped down to 5×10^{-6} mbar and then the ZnO source was sputtered for 10 minutes in the presence of argon gas with different pressures (0.01, 0.035, and 0.1 mbar) [55]. The growth duration was for 60 minutes and the rf power was 60 W.

As can be seen in Fig. 3.22., rather regular 1D nanowires have only been achieved in part (A). As for the other two parts, nonuniform randomly distributed nanostructures have been sputtered. It is also observed that the average diameter of the sputtered ZnO nanostructures is increasing with the increase of the argon gas pressure. For pressures of 0.01, 0.035, and 0.1 mbar the corresponding diameters were 124, 137, and 154 nm, respectively and the corresponding lengths were 960, 810, and 640 nm [55]. Also it is clearly seen that as the argon pressure is increased the grown ZnO nanostructures orientation starts to be random. Also as the pressure is increased the morphology starts to change from a nanowire to a flower-like morphology [55]. The axial or spatial growth in these experiments was explained as follows: the growth morphology and growth direction should in general depend on the growth conditions.

Figure 3.22 Scanning electron microscope images of ZnO nanostructures grown by the rf magnetron sputter deposition technique under various argon pressures (A) 0.01 mbar, (B) 0.035 mbar, and (C) 0.1 mbar [55].

The vertical axis growth at low pressures, that is, at 0.01 and 0.35 mbar, is justified by the relatively large mean free path of the adatoms with high migration lengths promoting axial growth leading to 1D nanowires by the side-wall surface diffusion along the direction of impingement [55]. These results have not been confirmed yet by other independent research efforts. On the other hand and for higher pressures leading to a shorter mean free path of the adatoms, these adatoms are most likely to be adsorbed on the side walls of the nanostructures, hence leading to the morphology observed in Fig. 3.22C [55]. This explanation is not convincing and is based on a weak argument, especially if we consider the special growth kinetics of ZnO (this will be discussed in more detail in a coming chapter). It is to be noted that the above explanation can hold for ZnO but cannot be a general explanation for the growth of materials in general, for example, for other metal oxides. It is to be noted that these results have not been repeated or confirmed by other researchers.

References

[1] M.J. Pitkethly, Nanotoday 3 (2003) 36.
[2] Available from: http://www.ssnano.com.
[3] S. Sivasankaran, A.S. Alaboodi, Functionalized nanomaterials. In: Muhammad Akhyar Farrukh (Ed.), Nanotechnology and Nanomaterials (2016) ISBN 978-953-51-2856-4, Print ISBN 978-953-51-2855-7.
[4] B. Zhang, R. Dugas, G. Rousse, P. Rozier, A.M. Abakumov, J.-M. Tarascon, Nat. Commun., Nat. Commun. 7 (2016) 10308.
[5] Y. Kim, K.-H. Ha, S.M. Oh, K.T. Lee, Chem. Eur. J. 20 (2014) 11980–11992.
[6] L.R. Harriot, Proc. IEEE 89 (2001) 366.
[7] Y. Todokoro, IEEE J. Solid State Circuits SC-15 (1980) 508.
[8] Z.G. Chiragwandi, O. Nur, M. Willander, N. Calander, Appl. Phys. Lett. 83 (2003) 5310.
[9] R. Hölzel, N. Calander, Z.G. Chiragwandi, M. Willander, F.F. Bier, Phys. Rev. Lett. 95 (2005) 128102.
[10] Z.G. Chiragwandi, O. Nur, M. Willander, I. Panas, Appl. Phys. Let. 87 (2005) 153109.
[11] Y. Xia, G.M. Whitesides, Angew. Chem. 37 (1998) 550.
[12] D. Qin, Y. Xia, M. Whitesides, Nat. Prot. 5 (2010) 491.
[13] P. Colson, C. Henrist, R. Cloots, J. Nanomater. 2013 (2013). ID 948510.
[14] P. Pieranski, Phys. Rev. Lett. 45 (1980) 569.
[15] U.C. Fischer, H.P. Zingsheim, J. Vac. Sci. Technol. 19 (1981) 881.
[16] H.W. Deckman, J.H. Dunsmuir, Appl. Phys. Lett. 41 (1982) 377.
[17] H.W. Deckman, J.H. Dunsmuir, J. Vac. Sci. Technol. B 1 (1983) 1109.
[18] J.C. Hulteen, R.P. Van Duyne, J. Vac. Sci. Technol. A 13 (1995) 1553
[19] G. Zhang, Dayang Wang, Chem. Asian J. 4 (2009) 236.
[20] J.G.C. Veinot, H. Yan, S.M. Smith, J. Cui, Q. Huang, T.J. Marks, Fabrication and properties of organic light-emitting "nanodiode" arrays, Nano Lett. 2 (2002) 333.
[21] P. Jiang, J.F. Bertone, K.S. Hwang, V.L. Colvin, Chem. Mater. 11 (1999) 2132.
[22] Z.-Z. Gu, A. Fujishima, O. Sato, Chem. Mater. 4 (2002) 760.
[23] D. Wang, H. Mçhwald, Adv. Mater. 16 (2004) 244.
[24] P. Jiang, M.J. Mcfarlan, J. Am. Chem. Soc. 127 (2005) 3710.
[25] F. Garcia-Santamaria, H.T. Miyazaki, A. Urquia, M. Ibisate, M. Belmonte, N. Shinya, et al., Adv. Mater. 14 (2001) 1144.
[26] G. Zhang, D. Wang, H. Mçhwald, Nano Lett. 7 (2007) 3410.
[27] D.M. Eigler, E.K. Schweizer, Nature 344 (1990) 524.
[28] L. Chen, X. Wei, X. Zhou, Z. Xie, K. Li, Q. Ruan, et al., Small 13 (2017) 1702003.
[29] R.S. Wagner, W.C. Ellis, J. Appl. Phys. 4 (1964) 88–91.
[30] Q.X. Zhao, M. Willander, R. Morjan, Q.H. Hu, E.E.B. Campbell, Appl. Phys. Lett. 83 (2003) 165.
[31] Y.C. Zhu, Y. Bando, Chem. Phys. Let. 377 (2003) 367.
[32] Z.R. Dai, Z.W. Pan, Z.L. Wang, Adv. Fun. Mater 13 (2003) 9.
[33] Y.C. Kong, D.P. Yu, B. Zhang, W. Fang, S.Q. Feng, Appl. Phys. Lett. 78 (2001) 407.

[34] O. Lourie, C.R. Jones, B.M. Bartlett, P.C. Gibbons, R.S. Ruoff, W.E. Buhro, Chem. Mater. 12 (2000) 1808.
[35] B. Xaing, P. Wang, X. Zhang, S.A. Dayeh, D.P.R. Aplin, C. Soci, et al., Nano Lett. 7 (2007) 323.
[36] S. Fujita, S.-W. Kim, M. Ueda, S. Fujita, J. Cryst. Growth 272 (2004) 138.
[37] A.Y. Cho, J.R. Arthur Jr, Prog. Solid State Chem. 10 (1975) 157.
[38] O.W. Kennedy, M.L. Coke, E.R. White, M.S.P. Shaffer, P.A. Warburton, Mater. Lett. 212 (2018) 51.
[39] S. Zhao, Z. Mi, Crystals 7 (2017) 268.
[40] K. Yamano, K. Kishino, H. Sekiguchi, T. Oto, A. Wakahara, Y. Kawakami, J. Cryst. Grow. 425 (2015) 316.
[41] B.H. Le, S. Zhao, Z.X. Liu, S.Y. Woo, G.A. Botton, Z. Mi, Adv. Mater. 28 (2016) 8446.
[42] A.K. Rishinaramangalam, M. Nami, M.N. Fairchild, D.M. Shima, G. Balakrishnan, S. Brueck, et al., Appl. Phys. Exp. 9 (2016) 032101.
[43] S. Zhao, H.P.T. Nguyen, M.G. Kibria, Z. Mi, Prog. Quant. Electron 44 (2015) 14.
[44] S. Zhao, A.T. Connie, M.H. Dastjerdi, X.H. Kong, Q. Wang, M. David, et al., Sci. Rep. 5 (2015) 8332.
[45] Q. Li, Y. Lin, J.R. Creighton, J.J. Figiel, G.T. Wang, Adv. Mater. 21 (2009) 2416.
[46] P. Dogan, O. Brandt, C. Pfüller, J. Lhnemann, U. Jahn, C. Roder, et al., Cryst. Growth Des. 11 (2011) 4257.
[47] D.B. Chrisey, G.K. Hubler (Eds.), Pulsed Laser Deposition of Thin Films, John Wiley & Sons, 1994. ISBN 0-471-59218-8.
[48] M.R. Rashidian Vaziri, J. Appl. Phys. 110 (2011) 043304.
[49] Nee, C.T., Tou, T.Y., Chapter 4 in Applications of laser ablation - thin film deposition, nanomaterial synthesis and surface modification, in: D. Yang, ISBN 978-953-51-2812-0, Print ISBN 978-953-51-2811-3, December 21, 2016 under CC BY 3.0 license.
[50] C.-V. Cojocaru, C. Harnagea, F. Rosei, A. Pignolet, M.A.F. van der Boogaart, J. Brugger, Appl. Phys. Lett. 86 (2005) 183107.
[51] W. Han, L. Jiang, X. Li, Q. Wang, J. Hu, Y. Lu, Sci. Rep. 7 (2017) 17333.
[52] R. Savu, E. Joanni, Scr. Mater. 81 (2006) 979.
[53] S. Choopun, N. Hongsith, E. Wongrat, T. Kamwanna, S. Singkarat, P. Mangkorntong, et al., J. Am. Ceram. Soc. 91 (2008) 174.
[54] M. Masłyk, M.A. Borysiewicz, M. Wzorek, T. Wojciechowski, M.E. Kwok, E. Kaminska, Appl. Sur. Sci. 389 (2016) 287.
[55] P.S. Venkatesh, V. Ramakrishnan, K. Jeganathan, Mater. Res. Bull. 48 (2013) 3811.

Further reading

C.L. Haynes, R.P. Van Duyne, J. Phys. Chem. B 105 (2001) 5599.

CHAPTER 4

New emerging nanofabrication methods

4.1 Introduction

Nanotechnology will clearly be the most dominating technology during the 21st century and it is the driving wagon for many new emerging applications in many different fields, like electronics, medicine, biotechnology, engineering, etc. Nanoscience is a challenging and interdisciplinary field that in many cases requires the knowledge of different branches of science. Nowadays, nanotechnology base projects and investigations are among the most funded worldwide. Hence, with all the allocated funds, the industry will demand well-characterized nanostructures with specific tuned physical and chemical properties. As the main driving force that led to the nanotechnology era is the need for smaller and handier electronic equipment, the main starting tools for dealing with smaller objects were the tools of microelectronics, i.e., you need a clean room to explore nanoobjects. Nevertheless, and due to the huge worldwide interest of scientists from different disciplines, the search for new desktop approaches for fabricating nanomaterials has continued. A quite large number of desktop-based techniques have been developed. Nowadays, it is not necessary to be in a complex laboratory environment like a clean room to perform research dealing with nanoscale objects. The terminology "nano" is now a common phrase among the public and is no longer reserved for scientists. This is despite the fact that an unjustified ongoing "paranoia" arguing falsely on the possible harm of nanomaterials to human health. Due to the huge seen and not yet seen benefits expected from nanotechnology, this false "paranoia" should be countered and defeated by providing the public with correct and solid information. Simply because if nanomaterials could harm our health, the human race would have suffered since its origins. As nanomaterials have existed in nature, bearing in mind that self-assembled (SA) nanostructures are the primary tool of the nature [1,2].

Low Temperature Chemical Nanofabrication
DOI: https://doi.org/10.1016/B978-0-12-813345-3.00004-6

This chapter is devoted to presenting some of the most common desktop techniques and approaches for the fabrication of nanoscale materials and devices for different applications. The recent development in newly adopted nanomaterial fabrication through synthesis, i.e., using the bottom-up route, has clearly indicated the fundamental importance of chemistry in the development of nanotechnology in general. This is evident from the fact that by using chemical routes, it is possible to start the nanofabrication by preparing specifically designed molecules or clusters that can act as nucleation sites for the growth of splendid nanostructured entities with specific tailored properties different from their bulk counterparts. Due to the importance of chemistry-based routes for the fabrication of nanomaterials, a separate chapter is devoted to identifying and discussing different chemical routes for nanofabrication. In Chapter 5: Low temperature chemical nanofabrication, the focus will be on the most commonly used approaches, choosing the most popular and well-developed approaches based on the number of published papers.

In 1854 and within the district of Soho, London, UK, and specifically in Broadwick Street, a severe cholera outbreak that ended the life of more than 600 people took place. Cholera as a disease was not known before this famous outbreak, although two much smaller earlier cholera outbreaks were reported in England. A physician named John Snow investigated this deadly outbreak. It was the discovery of the optical microscope that helped John Snow to identify the source that caused the Cholera outbreak. John Snow examined samples from the drinking water using an optical microscope, which was invented not long before that, and he was able to identify and confirm the presence of unknown bacteria in the water. Before that, society believed that all diseases were airborne. You cannot fabricate, use, identify, or deal with what you cannot see. Hence, we start this chapter with a brief review of nanostructures fabrication and imaging approaches; both standard as well as newly emerging techniques are to be presented. Bearing in mind that it is impossible to cover all emerging techniques with different approaches, we will present those techniques that are the most common or those with special features, e.g., high resolution.

4.2 Emerging top-down nanostructures fabrication and imaging technologies

As stated in Chapter 1, Introduction, the existence of nanostructures is probably as old as or even older than human existence on earth. But as mentioned earlier, you cannot use or control what you cannot see! By defining a nanomaterial as a material with one dimension with a size of 1–100 nm, it is very critical for the progress of nanosciences to develop reliable techniques for the fabrication and in situ imaging of three-dimensional nanostructures [3]. To fabricate a material the removal, addition, or displacement of a specific amount of this material is achieved through the supply of a certain amount of energy. The amount of material to be removed, displaced, or added, i.e., both the area and the volume, will largely depend on the energy that is localized in the specific location and the duration of the application of this energy. Photons, electrons, or ions usually carry this energy. Obviously, the resolution of the fabrication of a structure is related to the mechanism of supply of this energy. The ideal situation will be the development of high-resolution low-cost three-dimensional nanofabrication tools. Chemical, mechanical, and optical approaches have been widely used previously for microfabrication by drilling, chemical etching, lithography, etc. To improve the resolution limit of such approaches to the nanoscale domain, huge efforts have been devoted, for example, a laser (with short wavelength) is integrated into optical microscopy systems. In these efforts, photons with a wavelength λ can produce a diffraction spot size of $\lambda/2$ for durations of a few femtoseconds [3]. Indeed such efforts have resulted in the demonstration of subwavelength exotic nanostructures. Using such controlled energy delivery for fabrication has been successful in producing three-dimensional small-size materials [4–7]. Since atoms and molecules can have a possible complementary nature, e.g., shapes or charges, they can lock or attract each other. Then if millions of atoms or molecules are manipulated and brought together in a controlled way, then a nanoobject can be constructed. As of today, this has been achieved with different approaches. If this artificial object construction range is increased beyond a "nanoobject" to larger scales then a macroscopic object with a specific desired structure can be built, i.e., replicate objects artificially. On December 3, 1992, Neil Papworth, who is a test engineer working for Sema Group (now named Airwide Solutions), transmitted successfully the first "sms" message from his personal computer to the telephone of Richard Jarvis [8]. Before that date, nobody could imagine or believe that it is possible to send short text messages through computers or telephones. Although the idea of "molecular manufacturing or replicas" may sound like science fiction, it can one day in the future become a reality. The ultimate dream of nanoscience is to construct materials by starting from atoms to a specific required shape of a three-dimensional macroscopic nature, i.e., to obtain objects, starting from the atomic scale [4]. It is to be noted that some chemical-based routes, as can be seen in the next chapter, can perform this task, i.e., produce small-sized (a few nanometers) three-dimensional structures, and are desktop techniques with low cost, but with some challenging disadvantages. Below we will present some of the new emerging techniques based on the fabrication of nanosize structures using high-precision routes for deep sub-100 nm structures.

4.2.1 Ultrafast light-assisted nanofabrication

Most of the top-down high-resolution techniques including electron beam lithography (EBL) can provide resolution down to few nanometers. However, they cannot be used to produce three-dimensional arbitrary nanofabrications and obviously, the EBL is a relatively high-cost solution. In EBL systems, the operation voltage that accelerates electrons is usually in the range of a few keV and up to 125 keV with spot sizes down to 5 nm [9,10]. With such accelerating voltages, the electrons will have short de Broglie wavelengths (λ_B) given by $\lambda_B = h/p$, where h is Planck's constant, and $p = mv$ is the momentum of a particle with mass m traveling at speed of v [3]. Hence, short de Broglie wavelengths can lead to the achievement of high resolution, but focusing electron beams with such a short wavelength needs the use of complicated electrostatic and electromagnetic lenses [3]. This is contrary to the case of photons where focusing a beam of photons is easily achieved by focusing the photons to a focal point to achieve the limits of diffraction resolution [3,4].

In addition to the difficulty of focusing the beam of charged particles, energetic focused electrons or ions will cause local heating and electrostatic charging leading to damage of the underlying material. An example is the use of EBL in the fabrication of deep sub-100 nm metal oxide field effect transistor (MOSFETs). The steps that use EBL in the fabrication of ultrashort channel MOSFETs cannot be performed with beam energies higher than 10 keV at low exposure doses. Hence, the resolution will be compromised due to the relatively low accelerating energy and the low exposure dose will increase the exposure durations and lead to slow processing. This imposes limits to the downscaling trend for MOSFETs due to the reduced achievable resolution using EBL, due to restriction of the maximum allowed accelerating voltage leading to shorter wavelengths of the electron beam [11,12]. To overcome this challenging issue, researchers have proposed different solutions. Among them is the use of two-dimensional materials, e.g., graphene, however, it was found that electron beam irradiation will alter and modify the basic properties of the alternative materials used. To achieve a high brightness beam with temporal resolution other researchers have suggested the use of ultracold atoms to produce pulsed electron beams at extremely low temperatures. Despite the fact that these efforts will lead to complexity in instrumentation and increase the cost, continued research in this direction has been an ongoing activity [13,14]. Although developing such pulsed electron beams at extremely low temperature can improve the resolution and reduce the material damage, the concept of low-cost desktop three-dimensional nanofabrication will be completely lost. It is also to be noted that the challenge is not only on the underlying material to be patterned and on the possible damage caused by high-energy electron beams with relatively small spot sizes. The EBL process is based on the exposure of an electron sensitive material, i.e., electron sensitive resist. Hence the resolution is also dependent on the sensitivity of the resist to be patterned. Elastic and inelastic scattering interactions due to bombardment of high-energy electrons will be observed when EBL processing is used. Hence, both the primary and the secondary scattered electrons will travel inside the resist and will both contribute to the exposure process. This leads to the fact that the exposed area is larger than the beam spot size. Usually in EBL, the typical thickness of a high-resolution resist is smaller than the electron beam penetration depth [3]. As we discussed within Chapter 3, Conventional nanofabrication methods, when using the EBL, it is unavoidable to use multilayered resists stacked on top of each other. This multilayer-resist configuration is adopted as a solution to enable lift-off due to the restriction imposed with the relatively small thickness of the high-resolution resist. This multilayer resist has the high resolution deposited as the top resist to define the nano feature size pattern. This approach has mainly two disadvantages. It reduces the density of integration due to the possible large undercut and consequently restricts the minimum distance between adjacent nanosized objects, and will not allow closely adjacent long parallel contact lines.

Although the use of the multilayered resist configuration route in EBL has its drawbacks, it can be utilized to fabricate topographical shaped nanostructures, i.e., three-dimensional nanostructures [15]. Using multiple deposition layers of a high-resolution electron beam resist (EBR), e.g., hydrogen silsesquioxane (HSQ), three-dimensional nanostructures have been demonstrated [15]. The HSQ is an excellent high-resolution negative EBR with high aspect ratio.

As can be seen in Fig. 4.1, three-dimensional nanostructures have been fabricated by multistep EBL achieved through multilayer-resist deposition steps. The technique of multiple exposures using multilayer EBRs used for the fabrication of three-dimensional nanostructures is of interest for different applications that require periodic high-precision structures, e.g., photonic crystals, nanoelectromechanical systems, and Fresnel zone plates (FZP). An example of FZP fabricated using the multistep multilayer-resist EBL is shown in Fig. 4.1E and F. Such structures have the ability to increase focusing diffraction efficiency compared to conventional lenses. The HSQ resist is used for the fabrication of FZPs with higher pattern quality. Fig. 4.1E and F show a three-step structures suitable for FZPs with improved performance. FZPs fabricated using the HSQ are more efficient compared to previously fabricated FZPs as they can have a much smaller effective outermost zone with higher pattern quality [15]. Nevertheless, as can be seen from Fig. 4.1 the demonstrated topographical three-dimensional nanostructures, although are of a sizes suitable for FZPs, are not in the deep sub-100 nm. Despite the fact that EBL is a complex and expensive serial technique, it is widely used to fabricate high-resolution small-size patterns. Although it is not economically viable to utilize EBL for mass production, it is of potential for fundamental investigations to reveal the outstanding unique characteristics of small-size objects. Hence, it is important to investigate the minimum achievable feature size when using the EBL. A few factors limit the resolution of the EBL techniques. These factors are the minimum spot size of the electron beam, electron scattering, secondary electron kinetics, resist development, stability of the EBR, and the damage to the possible underlying material [3,16,17]. Almost all of these factors are dependent on the interaction of the EBR with the patterning beam. Hence, to improve and understand the limitations of the EBL technique it is important to quantify the energy deposited to the EBR at the nanometer scale [17]. This is achieved by measuring the so-called point-spreading function (PSF). Fig. 4.2 below shows a simplified example of the PSF effect [18]. Accurate measurement of the PSF is essential to quantify the overlap and hence determine the possible density of a nanometer-sized pattern (half-pitch) [19]. The electron energy loss spectroscopy technique can also be used to determine the pathway of the energy loss inside the EBR [17]. For improving the EBL resolution, transmission scanning electron microscope (STEM) was used as

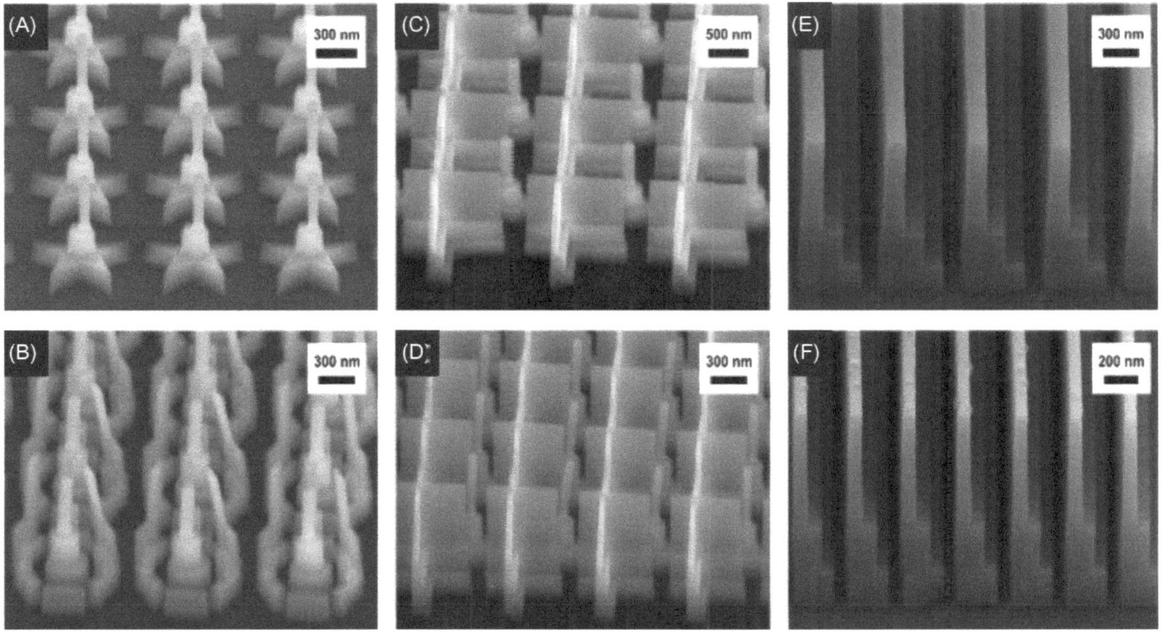

Figure 4.1 Scanning electron micrographs of three-dimensional nano structuring using HSQ negative mode resist. The final three-level patterns are the result of three consecutive overlayed EBL exposures. (A) and (B) show patterns having steps of about 100 nm. (C) and (D) Structures have lateral dimensions of 80 and 40 nm, respectively (E) and (F) Three-level gratings with periods of 600 and 400 nm using HSQ resist with a thickness of 1.2 μm [15]. *EBL*, Electron beam lithography; *HSQ*, hydrogen silsesquioxane.

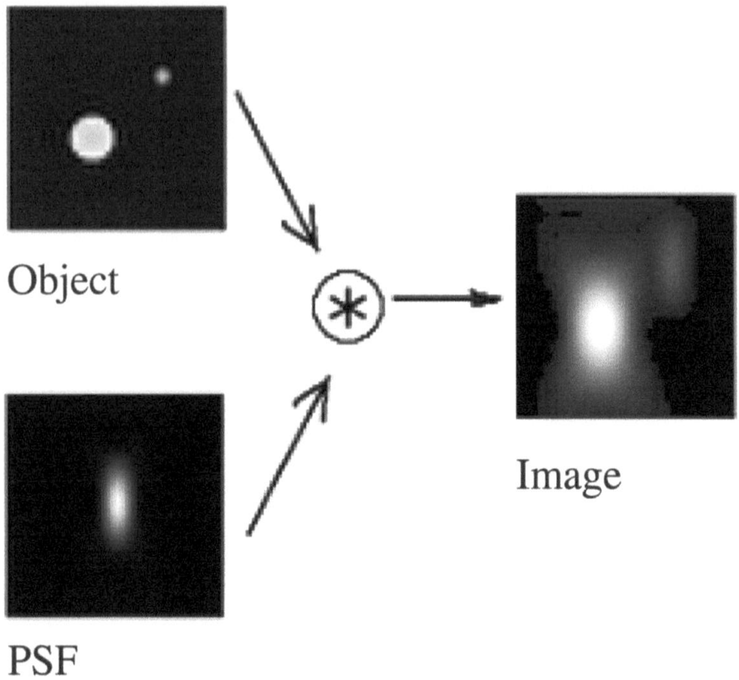

Figure 4.2 In general, the point spreading function (PSF) describes the response of an imaging system to a point source or point object. The figure shows a simple example of the effect of a PSF [18].

the tool of exposure. STEM can provide much higher beam energies compared to EBL systems, reaching up to 350 keV [17].

Using STEM, the resist interaction volume and the beam spot size can both be reduced. This will lead to improving the resolution to a few nanometers, as will be presented below. Different experiments on the use of STEM with high beam energies have been published, e.g., like those published in [20–22]. In those nanofabrication experiments, different types of EBR have been used. Although feature sizes down to 2 nm have been demonstrated, these attempts have some drawbacks. For example using low molecular weight EBR that is sublimated by electron irradiation like NaCl has resulted in feature sizes down to 2 nm, but excessive exposures, that is 500 times that needed for conventional resist, e.g., poly (methyl methacrylate), were necessary to achieve the relatively small feature size [20]. While the use of electron beam-induced deposition (EBID) has demonstrated a 2 nm feature size with 2 nm half-pitch, the time required for the EBID is 100–1000 times longer than conventional EBL [22]. These disadvantages render the approaches to be impractical and high-cost. Hence, such techniques are of interest for fundamental investigations, but not for mass production and industrialization, as it is a serial technique.

In an even more sophisticated method, an exposure using a dedicated aberration-corrected STEM, producing a minimum beam spot size down to 0.15 nm using 200 keV accelerating voltage, was used to identify the limits of EBL [17]. In this experiment, the minimum feature size, i.e., the resolution was determined by the forward scattering cross section of the 200 keV electrons, the secondary electrons kinetics, and the resist development chemistry for such relatively small patterns [17], i.e., the resolution in this case does not depend on the electron beam spot size. The resist used was the HSQ, which is one of the available EBR with highest resolution [23]. It is also to be noted that such high-resolution EBRs are quite expensive.

Some of the results of these experiments [17] are shown in Fig. 4.3. Fig. 4.3A shows schematic diagram of the processing steps. As the HSQ is a negative resist, SiN_x thin layer was used so as to avoid loss of resolution due to thick substrate etching, whereas in Fig. 4.3B and C the surface of the exposed substrate is shown after development of the HSQ. As can be seen no residues have been detected. One of the challenges in fabricating such small feature sizes are the adhesion of the resist and the collapse when the resist is developed. No such problems were observed in these experiments [17]. Fig. 4.3D shows thin lines of width 4 nm that were developed without any observed problem. Finally, Fig. 4.3E and F displays the 2 nm feature size achieved using a 200 keV accelerating voltage of the aberration-corrected STEM exposure. As can be noted one of the two nanoelectrodes has been shortened.

As mentioned above, the PSF is a key factor for understanding the limits of the EBL. In these experiments, the PSF was measured and the results were compared to the calculations [16]. The results of the PSF measurement for exposures with 200 and 30 keV are shown in Fig. 4.4. As can be clearly seen in Fig. 4.4A, the PSF at 200 keV is sharper than that of the 30 keV. This implies that the 200 keV electrons have a much smaller short-range proximity effect. Nevertheless, it was not possible to fabricate isolated islands with radius less than 2 nm. To estimate the PSF value for sub-2 nm for the H shape shown in the left insert of Fig. 4.4B, then a match of the critical dimensions and the radius of curvature of the

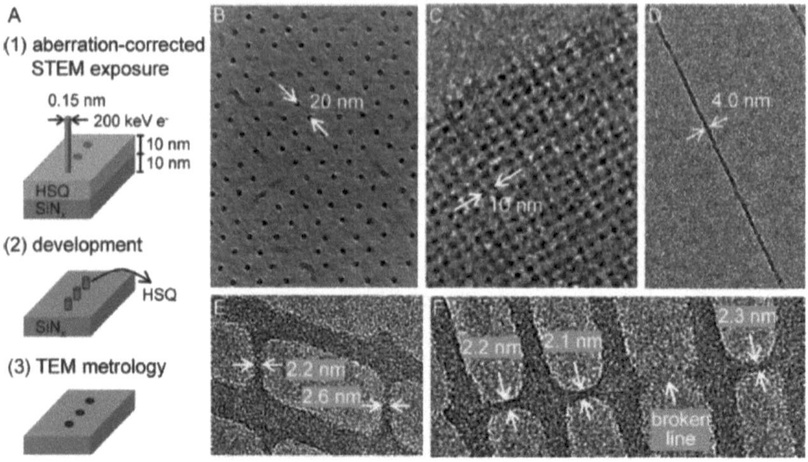

Figure 4.3 (A) Schematic of process using STEM, and TEM metrology using HSQ resist. The exposure was achieved using 200 keV creating 0.5–2.0 nm beam spot size with SiN_x membrane. (B) A 10 nm half-pitch HSQ dot array with 5.1 ± 0.8 nm average feature diameter. (C) A 5 nm half-pitch HSQ dot array with 5.6 ± 1.2 nm average feature diameter. The dose was (B) 18 and (C) 6 fC/dot (108,000 and 36,000 electrons/dot). (D) An isolated feature with average line width of 4 ± 0.8 nm. The linear dose was 21 nC/cm (14,000 electrons/nm). The diameter and line width variation represent one standard deviation. (E and F) The minimum feature size obtained by this method with features as small as 2 nm with a linear dose of 8 nC/cm (5300 electrons/nm) [17]. *HSQ*, Hydrogen silsesquioxane; *TEM*, transmission elector microscope.

H shape, and the density of the energy contour, the so-called calculated PSF was obtained [16].

In an attempt to overcome the limitations and challenges faced when using a high-energy electron beam as the patterning source, researchers have recently focused on the use of an optical beam as the means of direct writing for defining small-size (<100 nm) patterns. This has led to the so-called optical beam lithography (OBL), to replace the costly conventional EBL and avoid the challenges mentioned above, e.g., damage of the underlying material to be patterned [4]. Many researchers have demonstrated OBL attempts to fabricate deep sub-100 nm structures. The OBL is a rapidly developing field of research. However, the diffraction nature of light is a barrier for achieving nanometer features and hence limits the resolution in OBL. Below we discuss and describe some of the recent results demonstrating sub-100 nm structures using the newly developed OBL. One important criteria for a successful OBL approach is the cost. This is because even conventional optical lithography techniques used

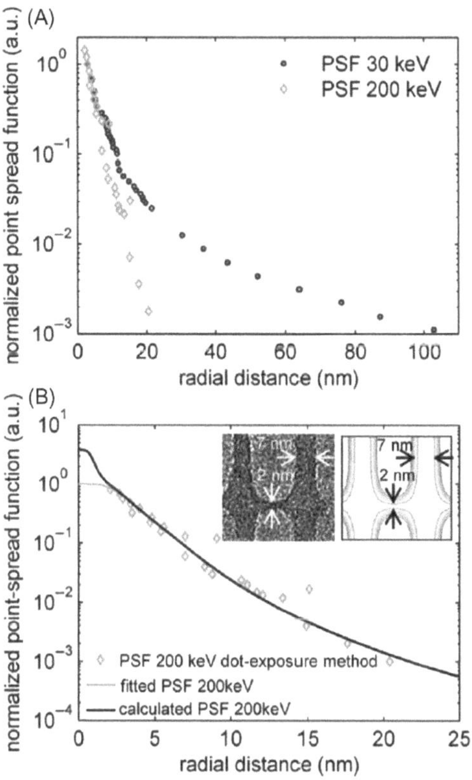

Figure 4.4 (A) The PSF for ~10 nm thick HSQ at 30 and 200 keV on SiN_x membrane substrate, showing that 200 keV electrons have a much narrower PSF compared to 30 keV. The PSF was measured by using single-pixel exposures. (B) The 200 keV PSF (*diamonds*) and the fitting function (lower curve at low radial distance curve). The PSF was iteratively calculated, named here as "calculated PSF," necessary to simulate the "H"-shaped test structure shown in the leftmost inset. The rightmost inset shows the resultant energy density contours using this calculated PSF. Specifically; the red contour in this inset defines the simulated structure using the calculated PSF. The calculated PSF (top curve at low radial distance) had an extra bell-shaped function with a knee at 1 nm radius [16]. *HSQ*, Hydrogen silsesquioxane.

today in fabricating sub-100 nm feature sizes in the microelectronic industry today are getting quite expensive. Today the cost of an advanced state-of-the art optical lithography tool can cost up to US$50M, and due to the process complexity, the mask alone can cost around US$5M due to the large amount of information needed to process a mask set which is usually processed using EBL [24]. Hence, developing a low cost maskless desktop OBL technique would be highly appreciated by the community.

However, the bottleneck will still be the low throughput of almost all maskless processes since they all are serial-based approaches. To overcome the serial process low throughput, multiaxes EBL systems have been developed and proposed to increase the throughput to allow an economically viable solution. Nevertheless, and due to the electrical charge interference causing coulomb interactions, and thermal drift, multielectron beam systems have suffered aberrations. Hence, it was challenging to control and regulate the beam spot size for such multiaxes systems [25–27]. In a different maskless process attempts, control over the resist kinetics to demonstrate subdiffraction small-size structures was achieved by assisted light beams [28–30]. Although the low cost criteria is satisfied when using assisted light beams, but the spatial regulation of the far-field optics affects the spot size [28–30]. As the previous discussion indicates, it is very critical when using a maskless approach to provide a mass production possibility as demanded by the industry.

A very critical issue in OBL for structuring materials is the development of a composite material suitable for polymerization (see discussion later). Development of near-field OBL systems and approaches has been another direction to overcome the limitations of conventional lithography. However, problems associated with difficulties in controlling the working distance between the optics and the substrate to be patterned and the control of the energy delivered have really hindered these attempts [31–33]. Further plasmonic lens (PL) has been suggested as a means to work out a near-field desktop mass production and low-cost approach to develop the so-called plasmonic nanolithography (PNL) [34]. Near-field plasmonic, nanofabrication approach has been popular and is developing.

In recent years, and due to the expected potential in the realization of mass production cost-effective deep sub-100 nm technology for nanofabrication, intensive research efforts has been performed in laboratories world-wide. However, before such approaches can be utilized for nanofabrication in the real world, some engineering challenges have to be solved. This is in addition to the need for efficient optical confinement schemes for the deep subwavelength to avoid the loss of surface plasmons, which will affect the throughput of the method. Below we discuss and present two examples of the successful use of PNL in realizing deep sub-100 nm fabrication. In the first investigation, a multistage PL capable of utilizing both propagating as well as surface plasmons, was used together with an air-bearing surface system. Using this system it was possible to efficiently squeeze light and demonstrate a 22 nm feature resolution using

355 nm pulsed laser light as the input source [35]. The design of the present focusing multistage optical system and its air-bearing system can be found in [34]. However, this system can improve the light transmission by 5–10 orders of magnitude. This will consequently ensure that enough energy is delivered for the direct writing of small-size patterns. Fig. 4.5 shows the plasmon lens, and the field distribution above the lens, together with a schematic diagram of the lithography exposure system with 10 nm distance between the lens and the surface of the resist. A key factor in this multistage plasmonic lithography system is the advanced air-bearing system that moves with a speed of 4–14 m/s at sub 10 nm

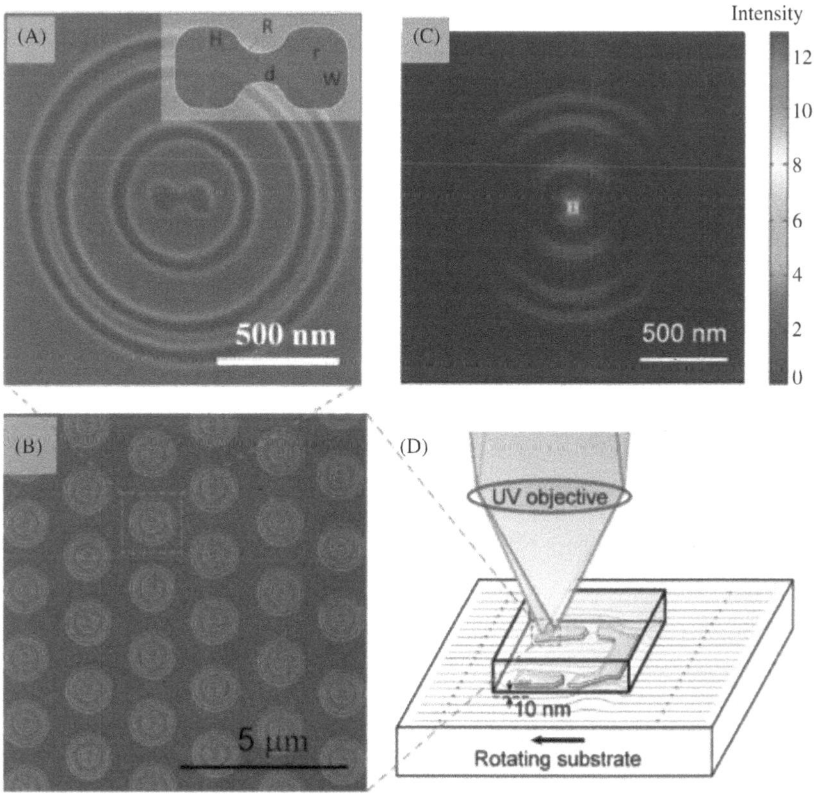

Figure 4.5 Scanning electron microscope image showing (A) the multistage plasmonic lens, (B) hexagonal multistage plasmonic lens array. (C) The field intensity distribution of a 355 nm UV light at a distance of 10 nm. (D) A schematic depiction of the multistage plasmonic technique with the lens having a 10 nm distance from the substrate at a linear scanning speed of 10 m/s. The distance is maintained by using an air-bearing technique [35].

distance above the sample surface. It is to be noted that the cost of advanced immersion optical lithography achieving a resolution down to the deep sub-100 nm by filling the gap between the last lens and the substrate by liquid Helium, is orders of magnitude higher than the present multistage PL-based lithography introduced above [35]. It is also of interest to mention that immersion lithography technology has been tested for 15 nm node technology, but as mentioned above the immersion optical lithography is an expensive technology. Some of the results of the multistage PNL exposure are shown in Fig. 4.6A–C. The displayed nanodots arrays (Fig. 4.6A) have a minimum feature size of 22 nm. These nanodots were obtained using a substrate velocity of about 7 m/s with a pulsed laser of 160 Hz [35]. Each single dot was generated by a single

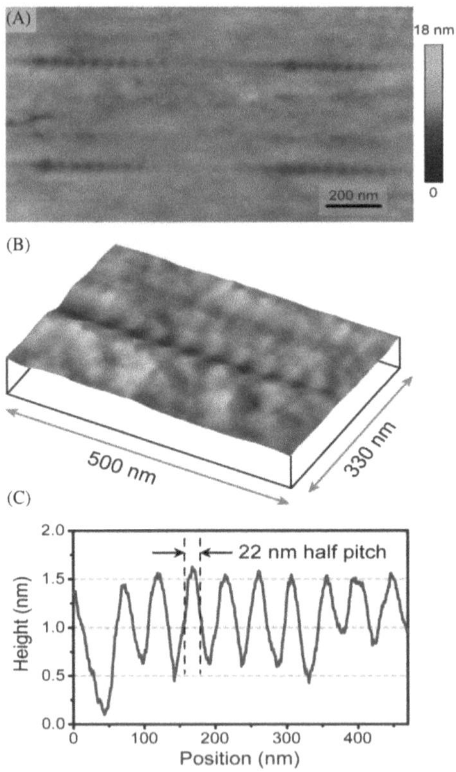

Figure 4.6 (A) AFM image of four groups of closely placed dots with a feature size of 22 nm obtained on a thermal resist. (B) Three-dimensional topography of the one dots line of part (A). (C) Cross-sectional profile of one of the dots line [35]. *AFM*, Atomic force microscope.

laser pulse exposure. By regulating the laser power, it was found that the exposed minimum feature size could be controlled. Further by optimizing the process parameters, e.g., resist exposure and postexposure processes, the pattern definition can be improved [35]. In Fig. 4.6B, a three-dimensional topography scan is displayed, while Fig. 4.6C shows a cross-sectional profile and I indicates that a 22 nm feature size has been realized [35]. It should also be noted that the pattern definition could be greatly improved by the optimization of the resist exposure threshold and preexposure steps. Although the above presented multistage PNL system is capable of defining deep sub-100 nm patterns, it cannot be utilized for fabricating three-dimensional nanostructures after exposure. Such three-dimensional deep subdiffraction OBL nanofabrication system has been developed recently [5].

Successful low-cost three-dimensional nanofabrication of well-defined deep sub-100 nm structures is essential and of interest for the development of future next-generation nanophotonic devices, e.g., nanowaveguides for maximizing light confinement, photonic crystals, ultrahigh density optical data storage, nanolasers [36–38]. As presented above (Fig. 4.1), it is possible to achieve three-dimensional nanofabrication using EBL (schematically shown in Fig. 4.7A) but with successive exposures and by using a multilayered resist (see Fig. 4.1) [15].

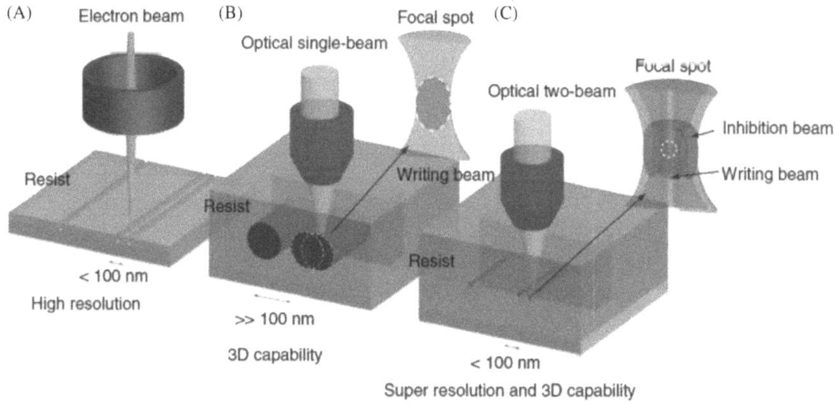

Figure 4.7 Schematic diagram for resolution and capability comparison between elector beam and optical beam high-resolution lithography systems. (A) EBL can routinely achieve 10–20 nm feature sizes. (B) Single-beam optical lithography system with capability of three-dimensional nanofabrication with limited resolution, and finally (C) two-beam optical lithography system with resolution comparable to that of the electron beam lithography [5]. *EBL*, Electron beam lithography.

To fabricate three-dimensional nanostructures, the use of the single-beam OBL approach with focusing aperture has been explored [5]. The obstacle in such a system is the diffraction limit imposing restrictions on the resolution leading to difficulty in achieving deep sub-100 nm feature sizes (depicted schematically in Fig. 4.7B) [5,39]. Further and by using two-beam OBL approach relying on polymerization and photoinhibition, the achievement of three-dimensional structures with deep sub-100 nm (similar to EBL) requires a special photoresist specification [29,30,40]. Such a photoresist should enable two-photon absorption, should have proper mechanical strength, and should have a specific photoinhibitation property. Such a photoresist will enable the nanofabrication of three-dimensional structures with high resolution, i.e., in the deep sub-100 nm feature size. With such a photoresist, a doughnut-shaped inhibition is achieved through the exposure beam (schematically depicted in Fig. 4.7C).

As mentioned above the key factor to demonstrate the three-dimensional subdiffraction OBL process, is the realization of a resin that possesses two chemical activation channels. The first channel is for photoinhibition, while the second is for photopolymerization [5]. Specifically [5]:

1. It should include an initiator with high sensitivity for two-photon absorption with minimum energy to solidify the resin to build structures with minimum feature sizes through photopolymerization sizes smaller than the focal spot of the writing beam.
2. It should have an effective inhibition of the two-photon polymerization process achieved through doughnut-shaped inhibited beam at a wavelength different from that of the exposing optical beam. With the condition that cross-excitation between the inhibiting and the excitation should be not exist.
3. It should have a minimum threshold regarding the energy needed for the two photopolymerizations to avoid possible material damage.
4. Finally, it should possess enough mechanical strength to withstand postexposure processing steps, i.e., developing, washing, etc.

The resin for this purpose should contain at least two components: the polymerizable material and an initiator [41]. The initiator or more specifically the photoinitiator will absorb the radiation and produce species that cause photopolymerization.

A resist that is suitable for the purpose of fabrication of three-dimensional deep sub-100 nm structures using OBL is 2,5-bis(*p*-dimethylaminocinn amylidene)-cyclopentanone (BDCC) [5]. This resin is capable

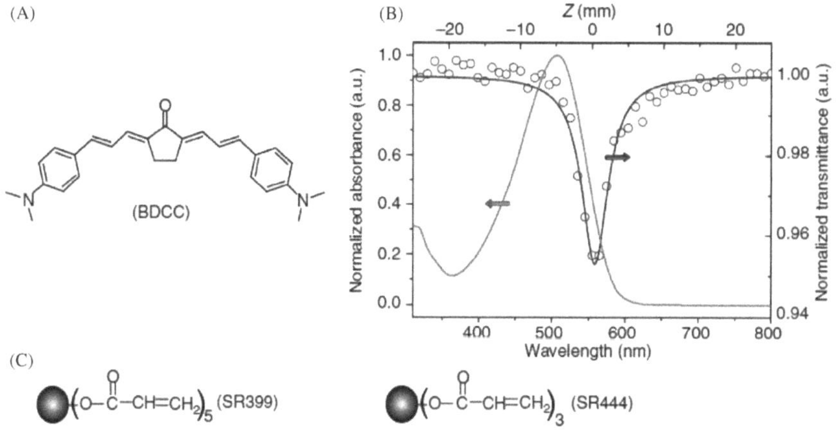

Figure 4.8 The structure and absorption of the BDCC photoresin. (A) The molecular structure of the BDCC. (B) Absorption spectra of the BDCC in chloroform showing single photon absorption and its two-photon absorption cross section. (C) Structure of the monomers with the BDCC, the SR399 for fast curing, and the SR444 with high viscosity to facilitate processing [5]. *BDCC*, 2,5-Bis(*p*-dimethylaminocinn amylidene)-cyclopentanone.

of satisfying the above four mentioned necessary requirements that enable the demonstration of three-dimensional deep sub-100 nm optical lithography. The structure of such resin is shown in Fig. 4.8. The details of the functionality of the initiator and the process of the two-photon absorption are complicated and can be found in [5].

Using the BDCC photoresin a three-dimensional deep subdiffraction OBL with a minimum feature size of 9 nm has recently been demonstrated [5]. The subdiffraction minimum feature size constitutes ($\lambda/42$) of the wavelength of the inhibition beam. In addition, a resolution of 52 nm ($\lambda/7$) two-line resolution for the photoresin with the ability to efficiently harness two-photon polymerization and single photoinhibition was also achieved. Using the BDCC photoresin described, the operation of OBL can be facilitated with threshold conditions that allow the reduction of the minimum feature size and allow the fabrication of dots and freestanding nanowires [5]. Fig. 4.9 shows some of the results achieved using the BDCC photoresin [5]. In Fig. 4.8 the two-line resolution concept was investigated. Two-line resolution is defined as the center-to-center distance between two parallel lines. Fabricating small-size deep sub-100 nm is necessary and enough to use as a prove of improved resolution, but the two-line resolution is also an important criteria and improvement of it is

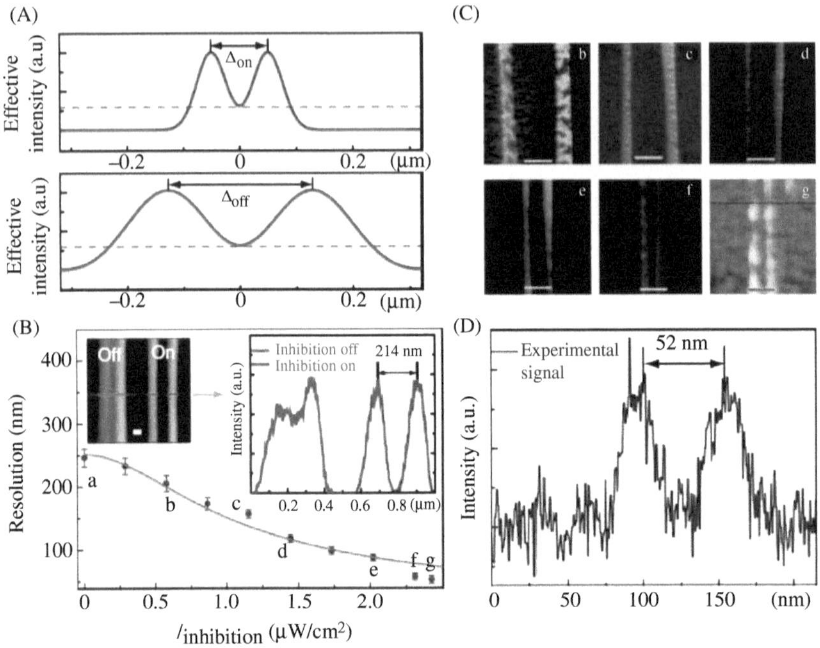

Figure 4.9 (A) Shows the improvement of the two-line resolution with the inhibition laser beam on (Δ_{on}) compared to the case when it is off (Δ_{off}), respectively. The horizontal dashed line indicates the threshold of the effective intensity. (B) The dependence of the two-line resolution on the inhibition beam energy. The insert shows two SEMs of two-line exposure with and without an inhibition beam, respectively, indicating clear resolution improvement with exposure including an inhibition beam. The error bars indicate the standard deviation. (C) SEM images for two-line resolution exposures using the parameters indicated by the points b–g, of part (B) of the figure. (D) Line profile showing the two-line resolution of the nanowires shown in g of part (C) of the figure. The scale bar indicates 100 nm.

of interest. Fig. 4.8A indicates the importance of the inhibition beam in improving the resolution of the two-beam optical lithography system. As can be seen in Fig. 4.9B, the two-line resolution is about 246 nm when the inhibition beam is absent (see point a in Fig. 4.9B). This 246 nm is almost the value of the diffraction limit of the 800 nm wavelength used for beam writing. By including the photoexcitation of enough of the photoinhibitors through the inhibition beam, the resolution limit can be improved. As can be seen in Fig. 4.9B, when the inhibition beam intensity is increased from 0.57 to 2.31 $\mu W/cm^2$, the resolution limit was observed to improve from 246 nm down to 57 nm [5]. As the inhibition energy is increased to 2.42 μW/cm^2, the two-line resolution has reached 52 nm. This value constitutes 1/7 of the used inhibition beam wavelength, it is also a five times improvement of the resolution compared to the case of exposure without the inhibition beam. Finally, in Fig. 4.9D, the line profile of the nanowires shown in part G of Fig. 4.9C, is shown indicating that 52 nm two-line resolution has been demonstrated [5]. This example [5] of the use of two-beam OBL indicated the viability of the approach to define subdiffraction feature sizes down to 9 nm, with two-line resolution of 53 nm. Freestanding nanowires have been demonstrated and hence, the approach has proven to be viable for three-dimensional nanofabrication using conventional laser sources. With more efforts, this technique has the potential to be developed to a technology that leads to a cost-effective portable system for nanofabrication based on OBL for deep subdiffraction feature size [5]. This development will be useful if the development of cost-effective resins is realized, something that is challenging.

Plasmonic nanostructures are among the category of materials that have been of interest in many research laboratories worldwide. This is due to their potential in producing many new prototype devices for applications in optical transmission, magnetic devices with response at the visible light wavelength, etc. [42,43]. Metallic nanoparticles are the main candidates for developing plasmonic nanostructures. The proposal of metallic nanoparticles as the potentially best candidates for developing plasmonic devices derives from their ability to provide and support resonant electron clouds at their surface, and due to their desirable response to electromagnetic waves at different wavelengths [44]. Due to this interest in developing robust and cost-effective methods, metallic nanostructures that are suitable for fabricating plasmonic particles have been an active research area. Although the fabrication of plasmonic nanostructures by dewetting of thin deposited films through thermal effects has been the most researched approach, thermal dewetting does not provide a convenient way of controlling the morphology of the produced nanostructures. Recently femtosecond laser beam as a tool of pattering thin films to demonstrate plasmonic structures has been proposed [44,45].

By using dewetting assisted by fs-laser pulses to gently remove material by successive multiple vertical pulse scanning, plasmonic nanostructures have been demonstrated [44]. In this investigation [45], a relatively large number of multiple pulses (about 10^4) are needed to achieve the demonstrated plasmonic nanoparticles. However, this method suffers from two problems. The first is the high degree of precision needed due to the relatively narrow laser pulse and the second is the relatively large number of pulses needed for this approach and these make it not practical for mass production. Recently a dual wavelength fs-laser beam with annular shape has been developed. This double peak annular shaped fs-laser has been utilized successfully for fabricating plasmonic metallic nanoparticles via dewetting of thin metallic films in a more convenient manner [44]. By applying the annular shape double fs-laser pulses to the surface of deposited thin films, a pattering of thin metallic films at the nanoscale is observed. The irradiation is then followed by ion-beam polishing for removal of the unwanted material and for further reduction of the size of the fabricated nanostructures. By adjusting the fs-laser irradiation parameters, it was possible to control the size, shape, and spatial distribution of the fabricated metallic nanostructures [44]. As shown in Fig. 4.10A and B, a single particle is formed. This is achieved for specific parameters of the irradiation [44]. Due to heating the gold (Au) is transformed to a nanostructure during the dewetting process by reducing the size of the central part, as shown in Fig. 4.10A and B. By focusing the beam using an objective of high magnification, a decrease of the size of the central pitch can be obtained. When applying the irradiation with the two-color dual wavelength fs-laser, the absorption by the film layer will be subjected to ultrafast heating effect with local melting. The substrate, being oxide or

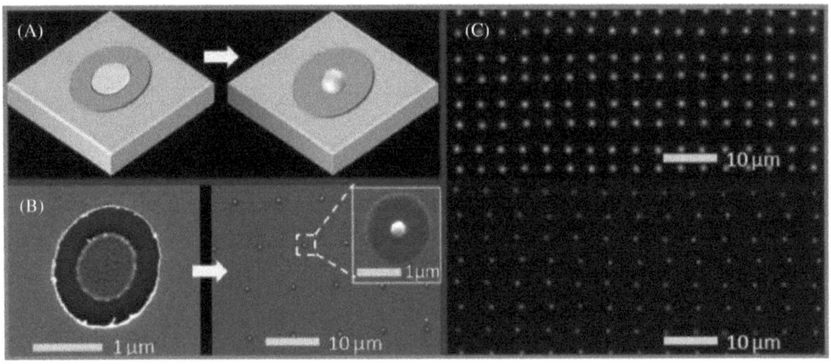

Figure 4.10 (A) Schematic diagram depicting single particle fabrication using dual wavelength fs-laser pulse from a gold film deposited on a dielectric substrate. (B) Scanning electron microscope images of concentric structures and gold thin film of 30 nm thickness nanoparticles using fs-laser dual wavelength irradiation with pulse energies of 0.012 and 0.024 μJ fabricated on silicon substrate. (C) SEM showing an array of gold nanostructures fabricated by the same method using pulse energies of were 0.021 and 0.026 μJ for the upper panel and lower panel, respectively [44].

semiconductor, and having a higher melting point, will remain unaffected. Due to melting, a mass reduction of the gold will exist. Fig. 4.10C shows the dewetting of a large number of Au nanostructures fabricated using this approach. The nanostructures shown in Fig. 4.10C were fabricated using a 40 nm thick Au film. As can be seen in Fig. 4.10C there are two panels. For the upper panel the energy used was 0.21 μJ, while it was 0.026 μJ for the lower panel [44]. As is clear, the fs-laser assisted dual wavelength approach is a lithography-free approach and with a single step nanostructures can be fabricated. It is useful for the fabrication of arbitrarily shaped plasmonic nanostructures of high quality.

4.2.2 Focused ion beam nanofabrication

Focused ion beam (FIB) was developed during the end of the 1970s and the beginning of the 1980s and was a successful application after the liquid metal ion sources were invented [46]. Although FIB was used initially for repairing masks, modifying electronic circuits to analyze failure causes, and for preparing transmission elector microscope (TEM) specimens, today its use has been increased to include four different areas of material science [47]. More recently, and with the development of high-precision FIB systems, the technique has been utilized for the fabrication of complicated nanostructures [48].

The FIB resembles the EBL with the advantages of being a maskless process operation with relatively higher current densities. In addition, FIB possesses the capability of much more precise focusing compared to the EBL technique. In addition, FIB benefits from the wide choice of possible different ions for operation, and because of the low current densities possible the penetration density in solids is shorter compared to EBL [49]. Another difference between the FIB and the EBL is the nature of the interaction between the irradiation beam being electrons or ions with solid materials. In the case of FIB, ions hitting a solid material can lead to displacement of the atoms of the targeted material, can cause surface sputtering, can also lead to the formation of defects, and will lead to other secondary processes [50,51]. This implies that when ions are bombarding a solid surface, effects like surface amorphization, deposition, swelling, and even nuclear reactions can all occur [47].

Fig. 4.11 shows the four basic applications of FIB technology. The complicated nature of the interaction between an ion and a solid leads to the unique capabilities possessed by the FIB technology. When an ion is

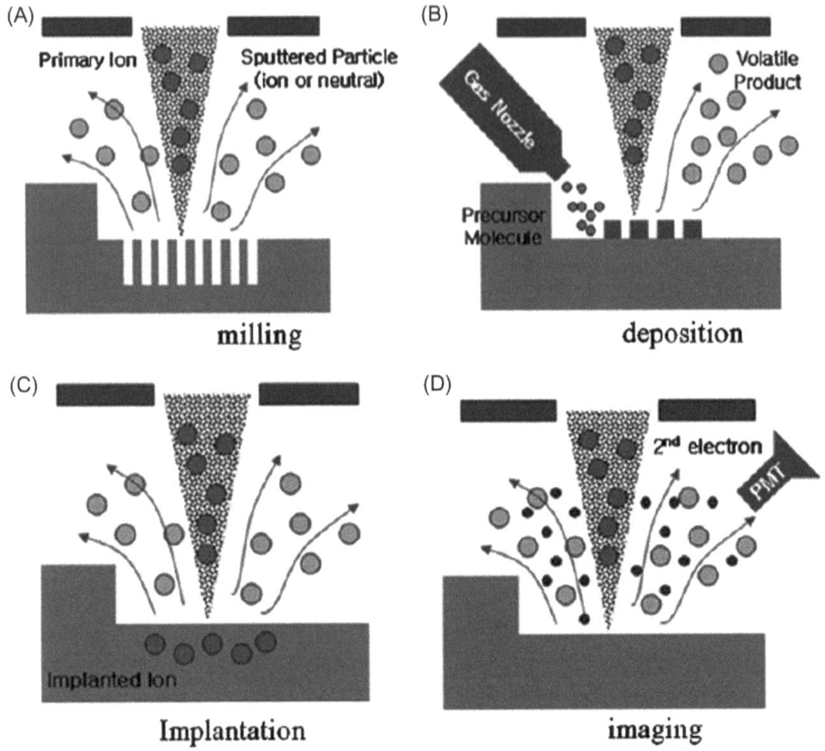

Figure 4.11 The four basic applications of focused ion beam system: (A) milling; (B) deposition; (C) implantation; and finally (D) imaging [47].

irradiated on the surface of a solid, the different main effects that can be observed are [48] (1) sputtering of surface atoms; (2) emission of electrons, which can be used for imaging; (3) crystal lattice atoms displacement accompanied by lattice damage; (4) photon emission; and finally (5) chemical reactions accompanied by gases and that can be utilized to facilitate film deposition. Hence and due to these effects, the FIB can be used to perform milling, deposition, implantation, and imaging, as shown schematically in Fig. 4.11A–D. Hence, the FIB constitutes a fabrication and characterization tool. The characterization is obviously through imaging, which is a very important characterization for nanomaterials [48]. To perform implantation (shown schematically in Fig. 4.11C) a certain specific mass of the ion source to be used and a specific accelerating energy are selected to enable penetration inside the solid target. For the imaging, the secondary electrons are collected by a photomultiplier while the ion beam is scanning a specific area of the solid target. In this way, imaging of the surface can be obtained. This is a very important advantage of the FIB chamber. It is important to mention that more than one of the four functions of the FIB can be utilized in one process. Further, the FIB can be used to perform cross-sectional milling by titling the sample. In turn, this facilitates cross-sectional imaging. This is of interest as it allows the imaging of the area below the surface of the target material. Although FIB can cause crystal damage, it has also been used frequently as an efficient route to prepare samples for TEM analysis [51,52]. Since the FIB is a high-precision tool, all the above four functions that can be performed by FIB are of interest for nanofabrication. However, deposition and removal of material are the most relevant for nanofabrication.

Hence, the understanding of the mechanisms associated with removing or adding material are essential for nanofabrication using the FIB as a tool. The addition or removal of material depends strongly on the ion source used, its ionic mass, the bombardment energy, and the nature of the solid to be removed or added [47]. Different ion sources have been developed for the purpose of the addition or removal of material, and information about these different sources is available in [50,53]. The surface atoms on the target material will be removed upon irradiation with ions emitted from the FIB source due to the transfer of energy due to bombardment. If the received energy is larger than the binding energy, the surface atoms will then be removed and milling or sputtering is then achieved. The sputtering yield, which is defined as the number of atoms removed from the surface of the target per incident ion is an important figure of merit [54]. The simulation of ion—solid interaction for a specific ion with a specific energy is important to acknowledge predicting proper removal of material for accurate fabrication of nanostructures. A software called Stopping and Range of ions in Matter (SRIM) has been developed and is widely used for the purpose of predicating ion—solid interaction for a wide range of energy and a variety of materials [55]. In the process of deposition of material, the target surface receiving energy less than the surface binding energy will not be removed (sputtered). However, it will rather be at an excited state at the surface. The excited surface atoms will relax and pass their extra energy to the adsorbed gas molecules [47]. Eventually these gas molecules will dissociate and will form a thin film on the surface of the targeted solid. Gas-assisted FIB for fabrication requires the understanding of the relationship between the gas dissociation and the replenishment and supply of the gas used [56]. Below we present some of the recent examples of the use of FIB in the fabrication of nanostructures.

Semiconductor patterns at the nanoscale can give rise to fundamentally new optical properties. In particular, regular complex nanoscale patterns can give rise to new optical properties, since they yield a high degree of optical chirality [57]. Such fundamentally new optical properties are observed due to the complex new unit cell shape, size, and alignment. Photonic crystals are a typical example of such complex well-arranged structures. The FIB and due to its precise control on the fabrication of nanostructures can be used to develop an unlimited complex arrangement of new shaped nanostructures that yield exotic optical properties [58]. The main parameters when performing nanofabrication using the FIB are the scan routine, dwell time, ion dose, and the angle of incidence. These parameters have to be precisely set to obtain the required shape and geometry of the nanostructure to be fabricated. The scan routine defines the type of scan used, i.e., raster, circular, .radial, or spiral. There are many different types of scan routes developed to facilitate the fabrication variety of structure in a convenient and optimum process. Details on the definition of these parameters and their significant effect on the fabricated structure can be found in [47].

Using a single crystalline silicon film deposited on sapphire, periodic complex nanostructures have been recently demonstrated [59]. The period of the fabricated structure was set to 370 nm to avoid visible light diffraction. The milling procedure was carefully controlled to achieve the highest regularity required for such chiral arrangement. All the parameters were set very carefully to maintain high quality beside the regular chiral arrangement. The area of the ion beam spot was chosen to be greater than the area of one pixel in order to achieve high fabrication quality [59]. A regular arrangement of chiral nanostructures was then periodically fabricated with an additional gap in between to minimize the influence of redeposition. The fabricated chiral structure is shown in Fig. 4.12. For each element in the chiral three-dimensional array, the fabrication started from the middle of the unit cell. Then milling continues toward the outsides. The array, which occupied an area of about 68 μm, was milled using an ion current of 0.1 nA and the milling time lasted for 2 hours [59]. After the fabrication of the three-dimensional chiral structure shown in Fig. 4.12A, the sample was first oxidized and then subjected to thermal annealing to reduce/remove possible surface defects that occurred during fabrication due to the bombardment of ions of the FIB.

To investigate the optical properties of the fabricated silicon chiral structure shown in Fig. 4.12, optical transmittance measurements were

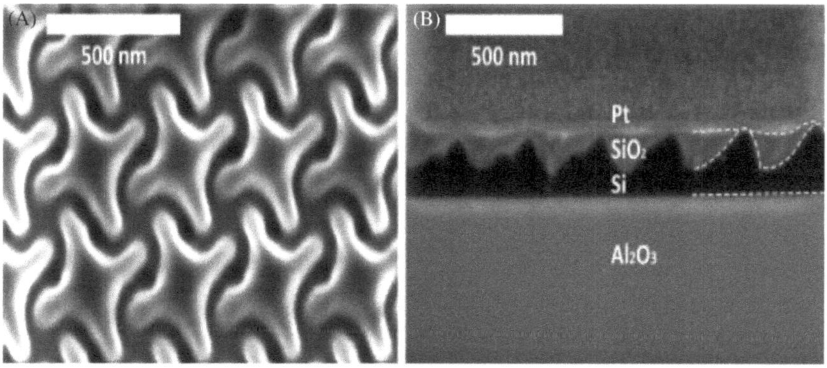

Figure 4.12 (A) SEM top view of the three-dimensional chiral structure fabricated by dual beam FIB and (B) cross-sectional view of the same structure in (A) after annealing [59]. *FIB*, Focused ion beam.

investigated. As can be clearly seen in Fig. 4.13, the chiral fabricated structure possesses higher visible transparency after fabrication/oxidation and annealing steps, compared to the flat silicon substrate [59]. This increase of transmittance is about 0.5 more compared to flat silicon for linearly as well as left (T_R) and right (T_L) polarized incident visible beam [59].

Although the nanofabrication by deposition and milling can be straightforward to understand, implantation can also be used to perform fabrication of different nanostructures [60]. Here we present one example of the use of implantation combined with other standard processes for the fabrication of nanostructures. Combining FIB ion implantation and wet-etching, silicon nanowires (NWs) have been fabricated and investigated for their piezoresistive effect [59].

The silicon nanowires were fabricated using silicon on insulator wafers with a top silicon layer having low doping (10^{14} cm^{-3}). After dicing the wafers into small stripes, a silicon amorphous NW embedded in the top of the silicon on the insulator was fabricated. This was achieved by gallium (Ga^+) implantation using a dose of 7×10^{15} cm^{-2}. Because the etching rate of amorphous silicon is slower than crystalline silicon, the silicon nanowires were obtained by an etching process. After Ga^+ implantation, which caused the implanted area to become amorphous, wet chemical etching was performed using tetramethylammonium hydroxide [59]. After amorphous silicon nanowires were defined, crystallization was achieved by annealing the nanowires at a temperature of 700°C for 30 minutes. Fig. 4.14A shows the fabrication steps and scanning electron microscope image of the silicon nanowire connected by two measurement pads.

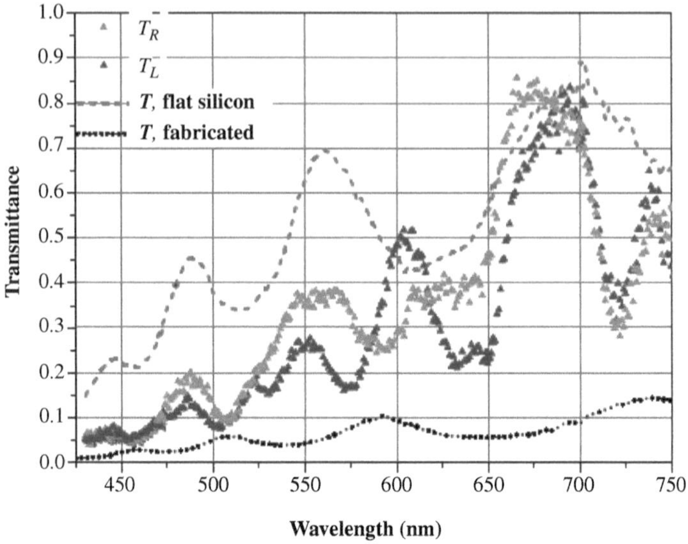

Figure 4.13 Optical transmittance of the FIB fabricated chiral silicon arrays. The case of flat SOS is also shown and represented by the dashed line. The chiral structure after the FIB fabrication is shown (*dotted line*), as well as the transmittance of the right (T_R), and left (T_L) circularly polarized light (*triangles*) of the annealed sample. *FIB*, Focused ion beam; *SOS*, silicon on sapphire.

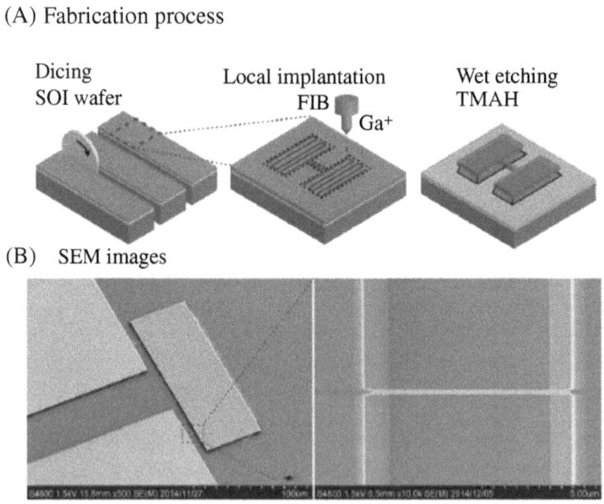

Figure 4.14 (A) Schematic representation of the fabrication steps. (B) Scanning electron microscope image showing the final silicon nanowires fabricated using the steps shown in 4.14(A) [59].

This example demonstrates that simple FIB implantation combined with standard wet chemical etching can be a route to fabricating nanostructures suitable for devices [59]. In this presented example, the interest was to investigate the piezoresistive effect of silicon nanowires. Recent other investigations of nanofabrication using the FIB as a tool can be found in [61–63]. Although successful utilization of the FIB as a tool has indicated success in the fabrication of nanostructures, there are many challenging issues to be resolved before the technique can be standardized for nanostructures fabrication. For example, milling can be accompanied by inherent redeposition in addition to other physical and chemical limitations [57].

4.2.3 Imaging of nanostructures

As mentioned above, you cannot use what you cannot see. This implies that imaging of nanoscale size objects is essential to fully utilize the exotic new properties expected from these small-size structures. Without proper imaging, we cannot relate a property to a specific structure and morphology. Due to this, the area of developing new tools and methods to visualize nanoobject has been part of the research on nanotechnology. The most basic tool used for imaging and visualizing nanostructures is the scanning electron microscope (SEM). The SEM is an electron microscope with its working principle based on focusing an electron beam that scans the surface of the sample. This electron beam will interact with the atoms of the surface of the sample producing different signals that contain information about the sample morphology and even composition. In SEM the electron beam scans in a raster manner. In high-resolution magnification, SEM can resolve objects down to 1 nm or even less. Nevertheless, as SEM depends on a high-energy electron beam, some disadvantages discussed before are also a problem when using the SEM as a tool for imaging in general.

Usually the FIB tool is also equipped with SEM for imaging; otherwise it is difficult to perform milling, deposition, or implantation. However, as electrons are charged articles, surface charging can lead to the loss of fidelity and can give false images causing misinterpretation. To avoid this charging effect, focused helium ions are used in some biological applications where cells and tissues will not withstand any charging effect, e.g., cell imaging [3]. Helium ions have a larger mass with shorter de Broglie wavelength compared to electrons. Such a short wavelength will enable the focusing of subnanometer beam spot sizes [3]. When a focused helium ion beam strikes the surface of the sample, it loses its energy through both elastic and inelastic collisions. In turn, this will lead to the generation of secondary electrons as well as ion-induced luminescence, and both of these effects can be utilized to perform high-resolution imaging [3]. As helium ions are positively charged, their charging effect can be neutralized through the application of electrons from another source in the system. These specific features of ion beams have been used to develop scanning transmission ion microscopy (STIM) that has been widely used to image biological small-size entities [64]. Helium ions are divided into two classes when considering them for imaging. First are slow helium ions, i.e., those having an energy of 10−100 keV. These slow helium ions can be focused to a spot of sizes of ~0.25 nm. Although such relatively small spot ion beam sizes of slow helium ions can be used to obtain ultrahigh superresolution, their disadvantage is that their penetration depth inside the material is relatively small [65]. There are also fast helium ion beams, i.e., having an energy of 1−2 MeV. With such high energy, the ions can pass through cells. The energy loss in this case will mainly be based on electronic ionization, with minimum scattering. However, fast helium ions can only be focused with spot sizes of the order of 30 nm. To achieve sub-30 nm imaging for biological applications, upconversion combined with ionoluminescence has been demonstrated [66].

By using incubating $NaYF_4$/Tm nanoparticles inside Hela cells, and applying a high-energy focused helium ion beam, upconversion is proposed and imaging beyond the diffraction limit is demonstrated [8]. Fig. 4.15 above shows the simplified experimental setup together with scanning electron microscopye of the nanoparticle and some of the results obtained using ionoluminescence compared to photoluminescence. As can be clearly seen in the insert of Fig. 4.15C, resolution enhancement has been achieved using ionoluminescence [66]. The achieved resolution using this approach was ~28 nm compared to ~253 nm when using a laser of wavelength 980 nm [3,66].

As imaging can reveal the exotic and interesting morphology related properties of materials at the nanoscale domain, scanning probes or focusing electron or ion beams have been the primary tools utilized. As it is clear from the published results, SEM is the most popular technique for imaging nanostructures. Utilizing field emission electron guns in scanning tools has led to resolution down to a few nanometers. However, and as

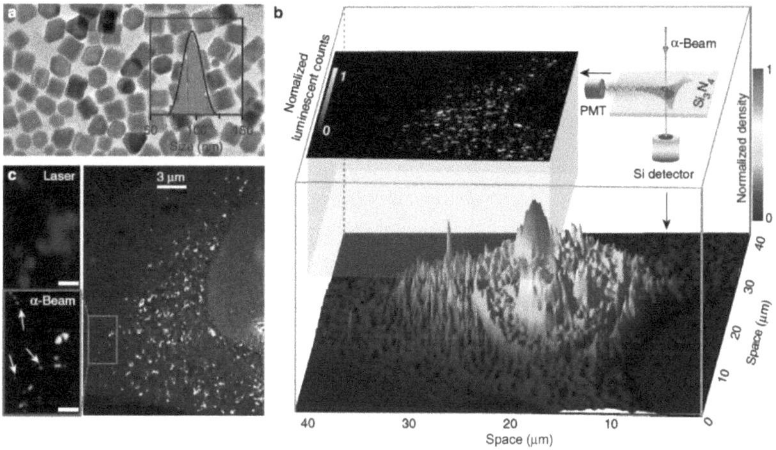

Figure 4.15 Example of ionoluminescence imaging for nanoparticles mapping. (A) TEM of the synthesized $NaYF_4$/Tm nanoparticles, the insert shows the size distribution of the nanoparticles. (B) Simplified schematic diagram of the experimental setup, ionoluminescence mapping of the nanoparticles incubated inside Hela cells after the nanoparticle uptake, and a three-dimensional rendering of the cell is shown. (C) The right shows the cells viewed through ionoluminescence and the insert shows imaging of incubated Hela cells with both photoluminescence (top insert) and ionoluminescence (bottom insert) clearly indicating resolution enhancement. *TEM*, Transmission elector microscope

mentioned above, electron beam bombardments will eventually lead to charging the surface of the sample under investigation. Hence when imaging insulating materials charging problems will cause damage and loss of imaging. In addition, in the case of three-dimensional nanoobjects distortion can be observed. These effects will lead to the loss of fidelity of imaging when using the SEM as a tool. To tackle these challenges, researchers have adopted the use of a thin conducting layer deposited on the samples to be viewed and the use of relatively small electron beam currents. The drawback of the use of a thin conducting layer is that it will mask the features of the surface of the sample, which is of interest. In addition, coating can be a destructive step that can render the sample useless after imaging. Relatively low electron beam currents lead to a long acquisition time to counter the diminished signal/noise ration. In turn, this will lead to a distorted image and the total loss of imaging fidelity. As SEM is a very convenient tool for imaging at the nanoscale and can provide high resolution; researchers have proposed other routes to tackle the disadvantages of the electrons' charging effect. In high-precision FIB, the

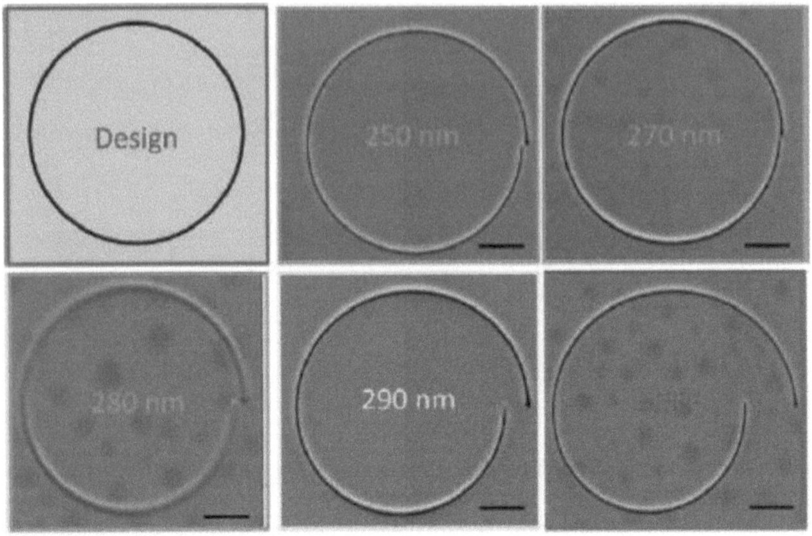

Figure 4.16 A scanning electron microscope of a circular pattern fabricated by FIB on a TiO_2 substrate and viewed under the illumination of UV with a different wavelength as shown in each figure. The scale bars represent 1 μm [67]. *FIB*, Focused ion beam.

photoelectric effect has been used to eliminate surface charging with deep ultraviolet coillumination beside the electron beam in SEM chambers. Energetic photons from the deep UV (260 nm wavelength) will have enough energy to liberate most of the electrons from the surface of the sample and lead to the reduction of the charging effect. In fact, the use of deep UV light will also affect the secondary electrons and make them experience a different electrostatic field and will lead to enhancement of the resolution. Fig. 4.16 below shows different SEM images of a circular pattern prepared by the FIB technique, while illuminating at the same time deep UV light with different wavelengths starting from 250 nm and up to 290 nm. Also in the figure, the original design as well as the milled circle with no deep UV light is also shown. The effect of charging is clear in the difference between the original design and the FIB fabricated pattern. The shown discrepancy between the design and the fabricated pattern is due to distortion caused by the charging effect that leads to the false image.

Absorption and scattering from nanotips have led to breakthroughs in imaging at tenth-of-nanometers to subwavelengths when combined with infrared and terahertz spectral bands [3,68,69]. Combining atomic force microscopy and infrared spectroscopy (AFM-IR) can lead to a useful tool for the chemical analysis of nanoscale objects [68]. The principle of the AFM-IR is as follows: the sample is usually place over an IR transparent prism, and then it is irradiated so total internal reflection occurs. Then the IR laser is tuned so it matches the absorption of the sample under investigation. Due to this match, the light absorbed by the sample will induce a photothermal effect. This will lead to rapid thermal expansion of the region that absorbed the photons. This rapid thermal expansion will induce a signal that is then detected by the AFM tip. The signal due to the thermal expansion will be proportional to the absorption coefficient of the sample. The detection of such a system (AFM-IR) is similar to photoacoustic spectroscopy, with the difference that here the AFM tip and cantilever are the detection elements and at the same time they are utilized as signal amplifying elements [68]. The utilization of the AFM tip to detect a thermal expansion pulse is the key factor that enables this technique to detect absorption below the diffraction limit. The AFM tip has the ability to sense and detect and map thermal expansion variation due to IR absorption for a spatial resolution lower than 100 nm, while it has a sensitivity that enables chemical identification for a scale of a tenth of a nanometer [68].

The scanning tunneling microscope is another tool suitable for imaging down to atomic levels, i.e., a resolution of a few angstroms. It was discovered in 1981 by Gerd Binnig and Heinrich Rohrer from IBM-Zurich, and they were awarded the Nobel Prize in 1988 [70]. The lateral resolution of the STM is down to 0.1 nm and the depth resolution is down to 0.01 nm [70,71]. Such resolution implies that atoms can be imaged and manipulated. Although the common use of the STM is under ultrahigh vacuum, it can be operated in air, water, and other liquids too. Further, the STM can be operated in a wide temperature range from near zero kelvin up to over 1000°C. The working operation of the STM is as follows: first a voltage is applied at the tip, then through a complicated control system the tip is brought in close vicinity to the sample surface, and when the vicinity distance is in the range of 0.4–0.7 nm, the tip maintains this vicinity. Such a vicinity is an equilibrium distance between attraction and equilibrium. At such a relatively short vicinity distance between the sample surface and the biased tip, electrons will tunnel between the tip and the sample. This will cause a current, which will then be measured. As the tunneling current starts to appear, the tip bias and relative distance between the tip and sample can be varied slightly.

As the tip moves along the $x-y$ plane, the variation of the surface and density of states will lead to changes in the tunneling current. The variation of the tunneling current or vertical distance z, can both be measured and will then be mapped. This mapping can be used to image the surface morphology. The STM can be used in two different modes, either constant height or constant current modes. For the constant current mode, as the height changes, the height is adjusted through a piezoelectric control system. The height variation will then lead to a topographic image of the surface. In constant height mode the bias is kept constant and the variation of the tunneling current is then measured. Hence a current image is then achieved, leading to charge density information. Although images obtained from the STM are gray, colors can be added to emphasize the features of the surface under investigation. The STM can also be used to obtain electronic information at a specific location by varying the applied bias and detecting the current variations. STM has been used to reveal the characteristics and nature of complicated interfaces, e.g., water molecule–solid interfaces [72]. Below we present one example of the use of STM to identify even buried thin layers through topography.

Relatively thin buried layers of dopants are crucial for the development of some devices, e.g., quantum emerging technology, advanced metal oxide semiconductor transistors, etc. These buried layers, i.e., delta doped layers or lattice mismatched crystals, can usually be characterized by destructive techniques or can be revealed from the final device characteristics through comparison with samples with no eta doped layer. Using STM operated at higher frequency can make such characterization possible with no need to destroy the sample under investigation. Scanning microwave microscopy (SMM) is another version of scanning tunneling microscope but operates at a higher frequency range (1–20 GHz) [73].

Using the SMM information, the complex impendence of buried nanostructures can be revealed and consequently other electrical information, such as conductivity and dielectric properties, can all be determined [73]. The experimental setup of the SMM is shown in Fig. 4.17. Topography images from silicon (0 0 1) with a thin buried delta phosphorus doped layer taken by atomic force microscope (AFM) and by SMM are shown in Fig. 4.18. As can clearly be seen, the AFM image did not yield any useful information about the buried layer, while the SMM topography image was successful in providing information of the spatial

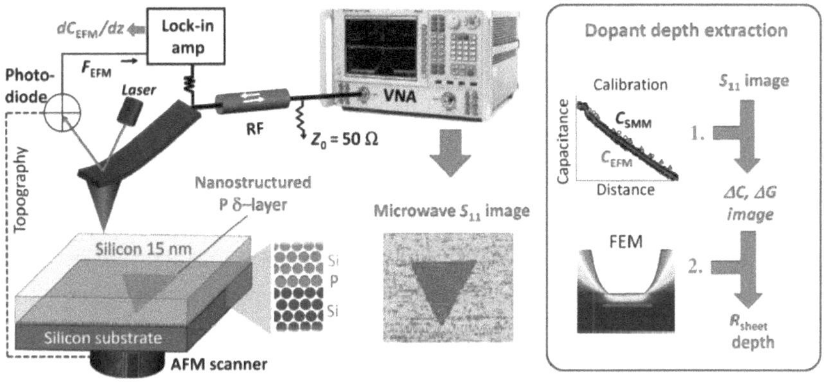

Figure 4.17 The scanning microwave microscope experimental setup [73].

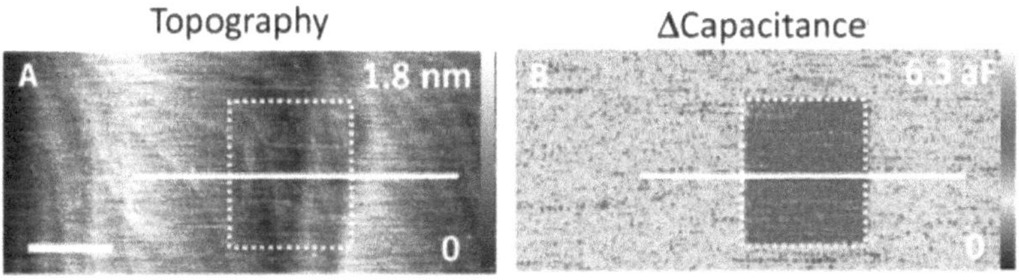

Figure 4.18 (A) Atomic force microscopy scanning of a silicon (0 0 1) surface with buried phosphorus delta doped layer. (B) The corresponding scan using a scanning microwave microscope showing a clear indication of the delta doped layer [73].

distribution of this layer as a different contrast indicated by the color difference shown in Fig. 4.18B [73]. The SMM provides a unique tool for characterizing buried doped nanostructures in a way that is not possible using other tools.

Although fundamentally complicated, the probing temperature of nanoscale structures can be a route for imaging. Because heat, unlike light, has a nonpropagative nature, the task of probing nanostructures using the heat emitted or absorbed by them is a complicated task. Attaining mature tools for thermal probing of nanostructures can be of interest for many applications, e.g., photothermal cancer therapy, drug uptake, nanosurgery, and nanochemistry. Despite the complicated issue of probing temperature for the purpose of nanoscale objects imaging, and due to its potential for the abovementioned applications, much research has been performed to develop a mature tool for this purpose. Different techniques for thermal imaging have been developed. The development of a hybrid sharp tip that is used as a nanothermocouple has paved the way for the start of thermal spectroscopy [74,75]. The first tool, an AFM-based technique, has demonstrated the capability of mapping temperature with a sensitivity of ~10 mK with a resolution of ~50 nm [75]. However, this early experiment relied on an invasive routine due to the proximity between the tip of the AFM and the sample to be mapped. Other less invasive techniques and tools relying on far-field optics for measurements and mapping of temperature have been developed in recent years, e.g., those found in [76–78]. However, most of these techniques have one or two drawbacks, e.g., lower sensitivity, slow readout, or the need for tagging the object to be imaged [79]. Recently another simpler thermal microscopy technique relying on sensitivity to thermal-induced refractive index variation has been developed. This thermal microscopy tool called TIQIS has an experimental setup that is much simpler compared to other tools as it combines Hartmann diffraction gating and a CCD camera. Although the resolution is 500 nm, i.e., not suitable for small-size nanostructures, compared to previous thermal microscope tools, it has advantages [79]. It is a fast technique and to obtain an image only one second is needed in most cases, there is no need for scanning over the sample, and hence there is no need for a complicated high-precision control system, and no modification of the samples to be imaged, e.g., adding fluorescent markers, is needed. In addition, the TIQSI provides two-dimensional images from a single measurement, and it provides quantitative information about the temperature, heat source density, and the absorption cross-section map for different parts of the nanostructure. This is in addition to other advantages over other previously developed thermal imaging nanostructures tools [79]. However, this tool, if developed for resolutions in the deep sub-100 nm, could be of potential for use as a widespread tool for imaging nanostructures.

Fig. 4.19 shows some of the results obtained using the TIQSI developed thermal imaging tool. Comparing parts (C) and (H) it is clear that besides temperature mapping, this tool can be used to design thermal nanostructures-based devices and configurations.

Another imaging technique, which is becoming very popular for imaging nanostructures in recent years, is coherent X-ray diffractive imaging (CXDI). As an imaging technique for nanostructures, CXDI owes its development to the demonstration of X-ray free electron lasers. The CXDI is a unique technique in the sense that an image is obtained with

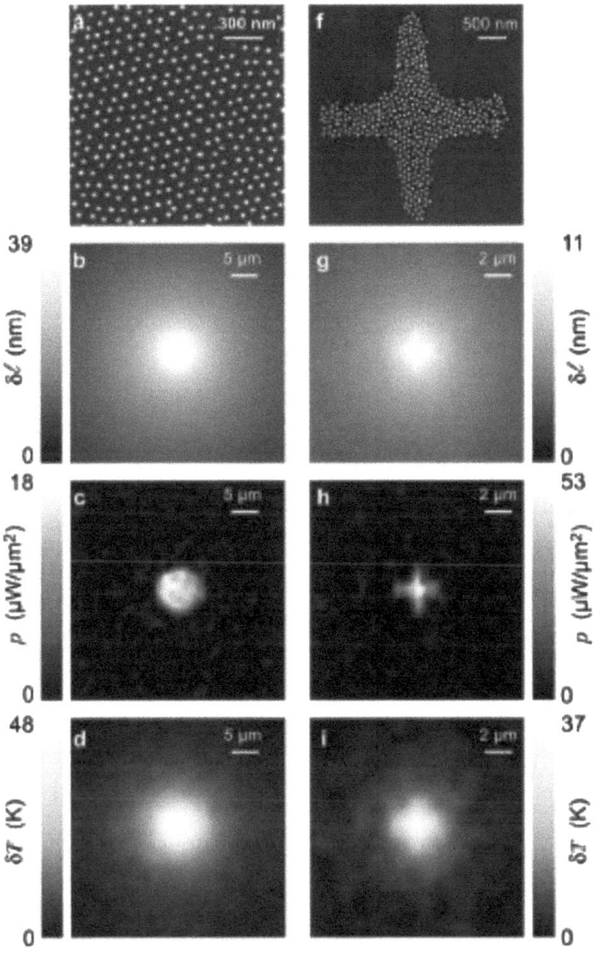

Figure 4.19 (A) Scanning electron microscope image showing hexagonal shaped gold nanoparticle arrangement. (B) Raw thermal image of the measurement of the optical path difference for the arrangement in (A) when irradiated by a circular beam with a diameter of 8.0 μm having an irradiance of 105 $\mu W/\mu m^2$ at $\lambda = 532$ nm. (C) The processed hear power density of (A). (D) The processed temperature variation. (F) Optical image of the whole hexagonal arrangement of the gold nanoparticles irradiated by a beam with irradiance of 250 $\mu W/\mu m^2$. (G) Raw thermal image of the optical path difference of (F). (H) The processed power intensity of part (F). (I) Heat source density thermal image [79].

no need for optics between the sample and the detection system [80]. This is schematically illustrated in Fig. 4.20. The basic principle of the CXDI is based on the fact that when a coherent light is reflected from a sample with a finite size a speckle is observed [81]. A speckle is observed

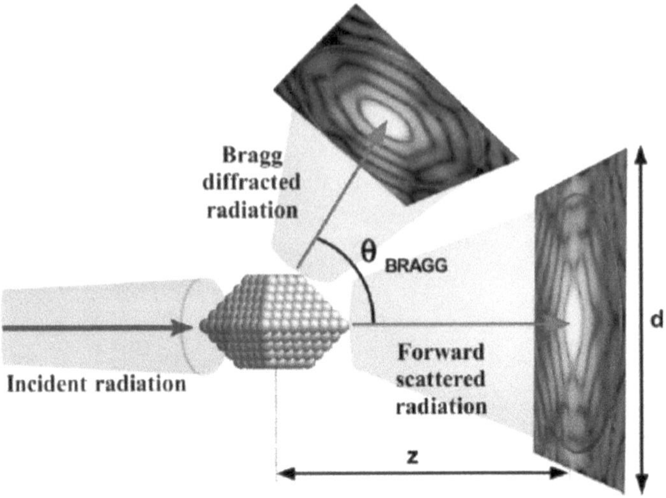

Figure 4.20 Schematic diagram representing coherent X-ray diffraction imaging with no need for optics between the object to be imaged and the detector. An incident coherent radiation falls on a sample with finite size and forward-scattered and diffracted radiation are far-field detected at a distance *z*. The detector can be placed either at the Bragg angle or at the downstream direction with regard to the sample [80].

usually when the material possesses regions of random distribution. The appearance of the speckle is due to the phase shift introduced in the scattered coherent incident radiation [81].

The intensity fluctuations in speckle pattern constitute a direct measure of the time correlation function of the inhomogeneity of the specific region of the speckle. The use of this effect by using visible light has been known for long time, and is called light-beating spectroscopy, dynamic light scattering, or, alternatively, fluctuation spectroscopy. However, due to the scale of the wavelength of visible light, it was not possible to probe or study lengths less than 200 nm [81]. However, with the advent of high brightness relatively short-wavelength lasers, the coherent X-ray imaging technique has emerged as a useful route for viewing nanostructures. Speckle from even single crystals has previously been observed using a coherent light with wavelength of ~0.15 nm [81]. The description of the basics of the CXDI and its different versions, especially for viewing finite size nanostructures is out of the scope of this chapter and can be found in [80,82]. Nevertheless, some important issues related to CXDI technique will be presented below followed by an example of imaging small-size nanostructures.

The kinematical X-ray scattering theory, which assumes that a single X-ray photon if scattered by an electron, will not suffer another scattering. This assumption is most likely to be satisfied by a relatively thin layer, and hence represents an acceptable approximation for the scattering of nanostructures. According to this theory, the amplitude of the scattered photons can be obtained as a Fourier transform of the electron density of the scattered atoms of the sample under investigation. In this case, the measured quantity will be the amplitude and not the phase, and hence the density of electrons will be ambiguously determined [80]. However, by using phase retrieval algorithms based on iterative methods, the structure of the sample can be revealed [80].

Although the coherent diffraction allows resolution in the sub-10 nm, the problem with conventional CXDI is that it can be used for imaging only very small objects. This leads to difficulty in the data quality and the type of sample that can be imaged. To overcome this difficulty, a new CXDI method has been developed. It is called the modified ptychography. On the other hand scanning transmission X-ray microscopy is another popular technique with the feature of straightforward data handling and analysis but the resolution is limited by the spot size; the modified ptychography combines both the coherent X-ray diffraction imaging and scanning transmission X-ray microscopy. Fig. 4.21 shows the experimental setup and an example of the diffraction patterns it is possible to obtain using the modified ptychography method.

Fig. 4.22 shows the raw data obtained from the modified ptychography, while Fig. 4.23 shows the processed data revealing all details of the

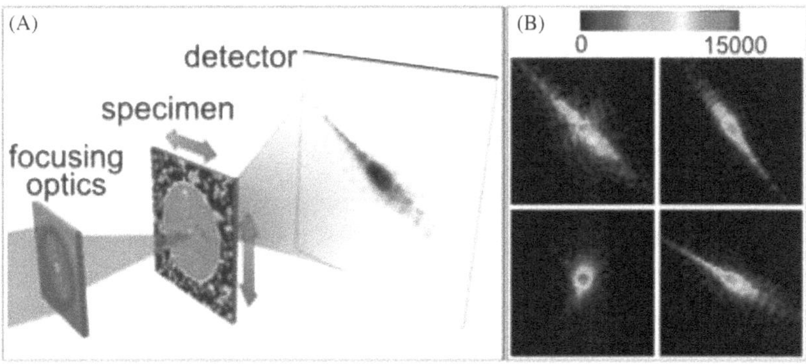

Figure 4.21 (A) Experimental setup of psychographic phase and amplitude reconstruction, and (B) example of the diffraction patterns [82].

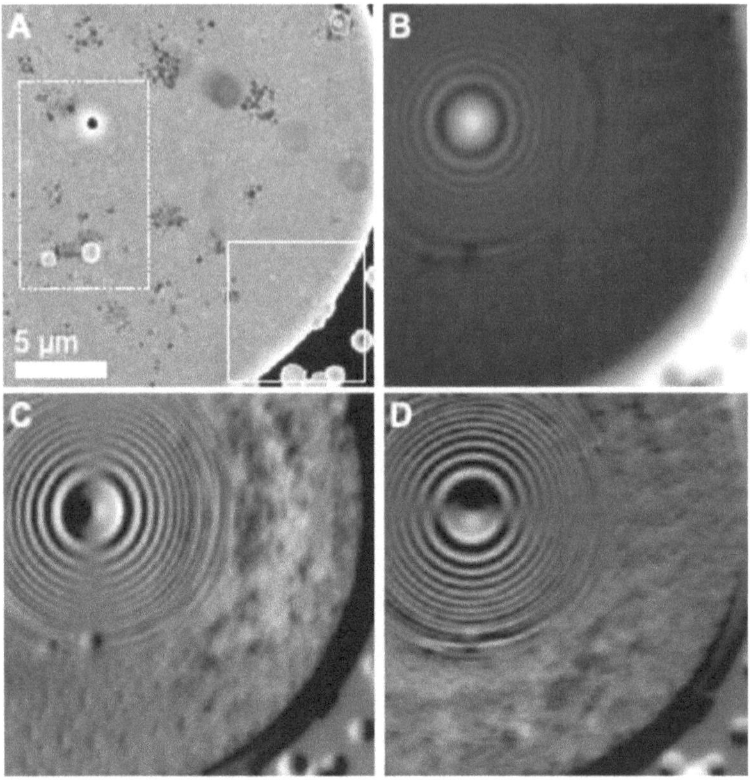

Figure 4.22 A preliminary data (raw data) for 201 × 201 diffraction patterns obtained from a Fresnel zone plate. SEM of a Fresnel zone plate having 30 nm diameter and 70 nm outer width. This plate is buried under a layer of gold. (B) Transmission intensity.(C) Phase gradient of the transmitted wave along the horizontal direction, and (D) Phase gradient along the vertical direction.

FZP imperfections. For more information on this topic and the specific details of these results, the reader is advised to read Ref. [82]. The modified ptychography method is a unique approach in the sense that it combines the high resolution of the coherent X-ray diffraction imaging and straightforward routines of the data handling of the scanning transmission X-ray microscopy. Further, the technique is noninvasive and the damage expected from radiation can be avoided by large area exposure and at the same time using high-resolution imaging for specific areas of interest. Both soft as well as hard X-rays can be utilized for this imaging, and the technique can be combined with other canning methods as described above to perform nanodiffraction mapping and nanospectroscopy [82].

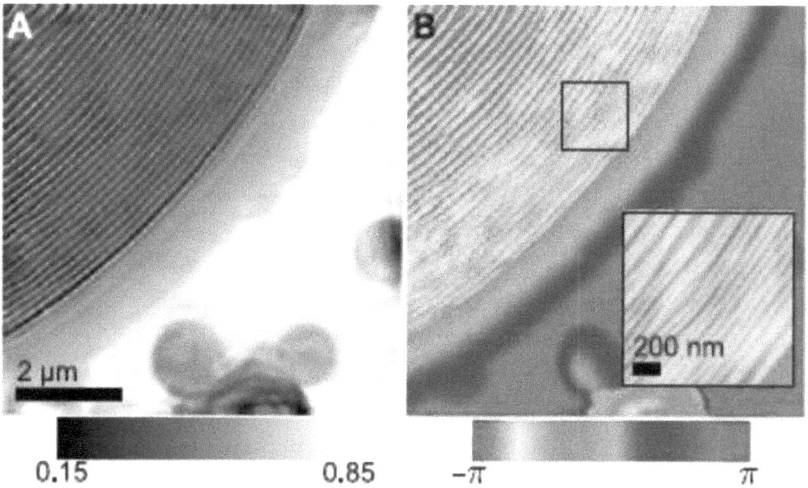

Figure 4.23 Scanning X-ray diffraction microscopy reconstruction of a specific region of 61 × 61 diffractions from the zone plate chosen form the data of Fig. 4.22. (A) Amplitude reconstruction, (B) Phase reconstruction. The insert shows the imperfection of the nanofabricated regions [82].

4.3 Self-assembly for nanostructures fabrication

Self-assembly is the tool of nature to create large systems from nanoentities. All organic or inorganic natural materials are fabricated/produced through self-assembly, starting from molecules or atom, sometimes even from electrons, protons, neutrons, or even smaller entities that are not known to us yet! This implies that nanoobjects have existed since long ago. In a self-assembly process small-size entities of a system assemble themselves in order to form a large well-defined long-range regular configuration. This action, i.e., spontaneous self-assembly, can be due to direct interaction between these small-size entities of the system or due to an external action from the environment around them. Through self-assembly very complex structures, leading to new materials of exotic properties, can be built. A stable SA complex system holds itself together at the molecular or atomic levels by intermolecular forces [83]. This means that the stability of the long-range SA system is due to characteristics coded as properties of the subentities. A stable self-assembly, i.e., a system at its final structure, implies that the system has reached equilibrium, meaning that it is at the lowest free energy state. In general, self-assembly is a process where minimization of repulsive force and maximization of attractive force between entities prevail [84]. Self-assembly is in fact a fabrication route that includes a lot of routes, e.g., in wet chemical reactions a synthesized material is due to random motion of molecules or ions with specific affinity and binding energy resulting in a self-assembly leading to the product. Self-assembly is now a well-utilized approach for so many disciplines and a variety of areas of research including electronics, biology, chemistry, engineering, etc. [85–87]. Nanoscience in general is the main benefactor of this natural fabrication tool as it can be used to build rather well organized long-order complex structures based on nanoentities. Self-assembly of molecules is today considered as an important bottom-up approach in nanotechnology.

The term "self-assembly" does not imply that a "random process" has led to the long arrangement leading to a larger system composed of small-size entities. In fact, self-assembly is governed by rules of thermodynamics. In a SA system, the entities of the system are at equilibrium in a state of lower free energy. SA system thermodynamics can be expressed by the Gibbs free energy of the SA system (G_{SA}). The thermodynamics in a SA system is given by the following energy equation [86]:

$$\Delta G_{SA} = \Delta H_{SA} - T\Delta S_{SA} \tag{4.1}$$

where ΔG_{SA} is the change of Gibbs free energy, ΔH_{SA} is the change of the enthalpy H_{SA} of the SA system, ΔS_{SA} is the change of the entropy of the system, and T is the absolute temperature of the system. The ΔH_{SA} is determined by the potential energy and intermolecular forces between the entities forming the system. A spontaneous SA can occur if ΔG_{SA} is negative. Usually self-assembly is observed with a decrease in the entropy. Hence, we expect spontaneous self-assembly when the enthalpy must be negative and larger than the entropy [88]. According to Eq. (4.1) as the magnitude of $T\Delta S_{SA}$ reaches the magnitude of ΔH_{SA}, and above a certain critical temperature, we do not expect spontaneous self-assembly to occur [88]. Nevertheless, still self-assembly exists under the conventional nucleation and consequently growth processes. Usually self-assembly of a system starts as a relatively "small" assembly. Since the lifetime of this small assembly will be increased due to the fact that attractive interactions between the different components of this small "assembly" will lower the Gibbs free energy. This will be followed by an increase of the volume of the assembly. Consequently and as the assembly grows bigger and bigger, the Gibbs free energy will continue to decrease and the lifetime of the growing assembly will also increase until the system becomes stable enough and can last for a long period. Hence, for a SA system to be in a thermodynamically equilibrium process, the existence of an assembly of a structure in a nonideal arrangement must exist followed by lowering the Gibbs free energy of the system to a minimum value that keeps this nonideal arrangement stable for a long duration.

Most of the demonstrated self-assembly complex nanostructures have nanoparticles as the individual constituents. The processes by which these nanoparticles assemble themselves in a complex long-range configuration are many and depend on each individual case. Understanding the nature of the process of an individual SA configuration is critical for the optimization of the configuration itself. However, these processes and effects that lead to the spontaneous self-assembly are related to information coded in the memory of these nanoparticles. In general, nanoparticles assemble themselves for two broad reasons [89]. These are (1) molecular interactions, or (2) external effects. The self-assembly of nanoparticles can prevail through the formation of covalent or noncovalent bonding. The bonding of neighboring nanoparticles are responsible for the self-assembling and they are called capping ligands [89]. By careful design of ligands as building blocks, exotic new "nano" systems can be developed, e.g., nanosensors, nanomachines, nanorobots, etc. [90]. External effects leading to spontaneous self-assembly are many, e.g., temperature, pH, and external electric or magnetic fields, to mention a few. Below we present some of the recent demonstrated complex nanoconfigurations achieved through well-designed self-assembly of nanoentities.

Spontaneous self-assembly of nanoparticles can be achieved in solutions. In this case, there is no need for a template, or the need for an interface or external triggering action to achieve the task of self-assembly. One possibility to achieve a SA freestanding complex configuration of heterogeneous nanoparticles in liquids is to use their end bond site–specific chemical properties [91]. In addition, self-assembly can also be achieved at interfaces of other solid materials [85]. The effects to trigger the self-assembly in solutions are many, but all are related, as mentioned above, to information coded in the individual entities' memory or by functionalizing their end site groups [92]. For the self-assembly through the functionalization of a template, a variety of materials, such as viruses, DNA molecules, copolymers, carbon nanotubes, etc., can all be utilized. The nanoparticles/template interaction is the basis that leads to the self-assembly of the nanoparticles. This is illustrated schematically in Fig. 4.24A above. Fig. 4.24B shows an example of anionic poly(poly(vinylpyrrolidone))-functionalized gold nanoparticles SA on the surface of

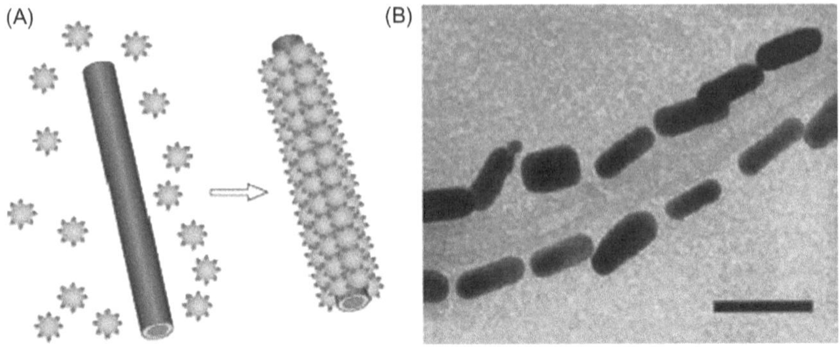

Figure 4.24 (A) Schematic diagram illustrating the attachment of nanoparticles to the surface of a cylindrical, and (B) transmission electron microscope image of gold nanorods attached to a tobacco mosaic virus (scale bar represent 50 nm) [93].

a carbon nanotube [93]. Nevertheless, the use of hard templates, e.g., carbon nanotubes, has the drawback of lacking the control of the spacing between the SA nanoparticles. This can clearly be seen in Fig. 4.24B where the space between the functionalized gold nanoparticles is irregular.

Since self-assembly as a tool for nanofabrication provides a huge potential for many branches of nanotechnology. Especially the area of nanomedicine can benefit a lot from self-assembly and interesting applications has already demonstrated. It is interesting to note that the concept of self-assembly was originally inspired by viral biology [94]. This is very evident from the fact that the biological systems are the most complex regular SA systems. In fact, biological systems have always been a source of inspiration for nanotechnology. For example, the cell, which is the basic fundamental unit of life, is composed of different separate nanosystems that separately or jointly perform very complicated functions [95]. These different nanosystems interact with each other through very complicated chemical pathways that we know very little about. Further, these cellular nanosystems self-assemble themselves in a complex regular hierarchical way generating complicated structures that sense and regulate the environment around them. In general, cellular nanosystems perform and enable many complicated functions like replication, motility, apoptosis, etc. [95]. In-depth complete knowledge of the chemistry and physics of the cell is in fact a dream of the scientific world. With the advances of nanoscience, and the development of tools to characterize and mimic cell conditions, this dream could probably one day become a reality!

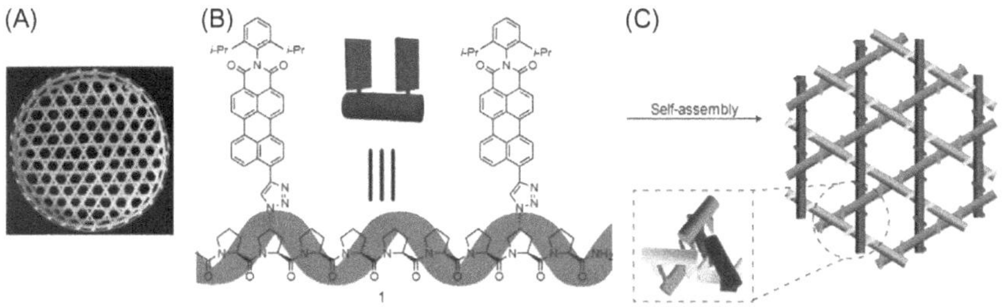

Figure 4.25 Illustration of adopting the weaving using molecules. (A) A picture of a basket woven in a triaxial pattern' (B) Molecular structure of the molecules used to form the artificial woven configuration' (C) Top view of the woven molecules forming the triaxial structure.

Self-assembly has been applicable in many other areas inspired by different special properties of certain configurations. For example, weaving, which is a very old technology where directional threading can lead producing a material much stronger than the original threads, has inspired scientists to use self-assembly to replicate weaving at the nanoscale. Indeed the successful demonstration of nanowoven structures has led to new exotic structures. Chemists have recently developed methods to weave molecules, forming complex patterns of exotic properties. An example is a recent demonstration of a triaxial weave that has been fabricated using the self-assembly of well-designed organic molecules through bonding interactions. This example is shown in Fig. 4.25 [96].

A research area that has been inspired by and benefited a lot from the replication of weaving using molecules is the DNA nanotechnology research area. Since it was proposed in 1981, DNA nanotechnology has developed a lot [97,98]. The original idea proposed the use of branched DNA junctions to form two- and three-dimensional crystal lattices to act as hosts for many purposes. The aim of the DNA nanotechnology via self-assembly is to precisely design functional molecules that are not possible to demonstrate using conventional crystallization methods that can perform in vivo therapeutic drug delivery [98,99]. Indeed, through rational design of DNA nanotechnology using self-assembly, many macroscopic crystals for attaining specific functions like proteins, peptides, and other nanoparticles have been demonstrated [99].

As it is evident from the above discussion and examples, self-assembly provides a desktop simple and low-cost method for producing ensembles of nanoparticles in a controllable manner. To achieve successful SA configuration of nanosized entities, extended interactive connections between the

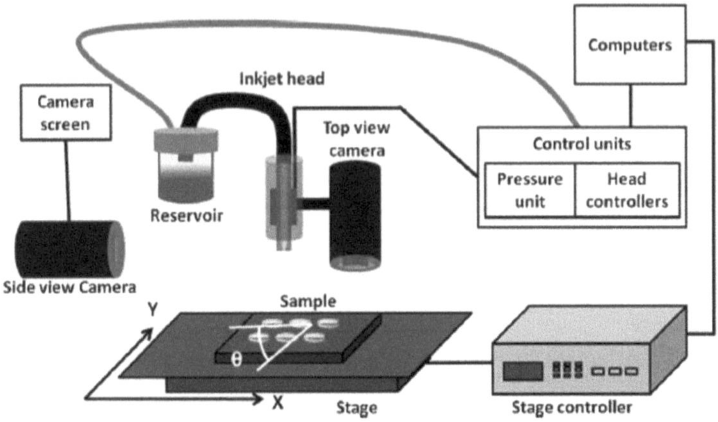

Figure 4.26 Schematic diagram of a typical inkjet printer. The sample is mounted to a computer-controlled stage. The ink solution is pushed through pressure to the nozzle tip. The drop generation is controlled via a piezoelement. The alignment of the inkjet head with the substrate is adjusted through a camera.

individual constituents is necessary. Indeed many molecules satisfy this requirement. In most of the conventional materials we deal with, e.g., polymer, gels, and liquid crystals, the structural configuration site end groups and their motifs are quite simple compared to other natural molecular biological structures with highly hierarchical complex arrangements [89]. This simply implies that researchers adopting self-assembly as a tool for nanofabrication for constructing complex multifunctional materials and to facilitate smart systems should learn from natural biological systems. This is due to the fact that the biological natural systems are the most sophisticated hierarchal complex structures known to humans up to now! Hence, we need to mimic the fabrication routes of nature in order to fully utilize self-assembly to the maximum, but it is not an easy task, as we still need to know more about the fundamental issues in that respect. Although self-assembly is a bottom-up approach, combining it with state-of-the-art top-down routines for nanofabrication would enable the demonstration of relatively advanced exotic applications.

4.4 inkjet printing as a tool for nanofabrication

inkjet printing is an emerging desktop nanofabrication method that is low cost with high throughput [100]. In an inkjet process, material is ejected from a nozzle and deposited on a substrate with a specific color and pattern depending on the ink nature, as depicted in Fig. 4.26. The original idea of inkjet dates back to 1878 when it was shown that a liquid could break into smaller droplets [101,102]. IBM developed the first inkjet printer in 1970s.

Although the technique of inkjet was originally for printing, it is now realized to deposit nanostructures for many applications, e.g., electronic printing, solar cells, light-emitting diodes, biosensors, optical materials [103]. As a rule to deposit functional nanomaterials the inkjet requires the prepreparation of a proper ink. Once a proper ink is prepared, injection can be performed. The inkjet printing has been extended by assistance with other processing tools, e.g., UV pulsed laser irradiation or a prepre pared patterned substrate [100,104]. One advantage of inkjet printing is that it can allow the use of any substrate and hence could enable an economical mass production route for nanostructures-based paper electronics. Recently inkjet printing was used to demonstrate a scalable three-dimensional nanostructured graphene on rigid as well as on flexible paper substrates [105]. Below we present some of the recent examples of depositing nanostructures on different substrates.

As the interest in disposable paper electronics is increasing, new innovative methods are of interest to be developed. One approach is to use inkjet printing, as it is compatible with paper and can be scaled up for mass production to meet economic viability. Indeed recently processing of silver nanoink has been demonstrated as a route to deposit highly conductive thin layers [106]. Fig. 4.27 shows the steps of the adopted process together with SEM images of the produced silver nanoparticles. To produce regular silver nanoparticles with good conducting properties, postprocessing multisteps have been adopted as described in [106]. Colored printing has also benefited from the use of nanoparticles or the interference of nanostructures to produce beautiful photographs. An example of this can be found in reference [107]. Fig. 4.28 displays an image produced using inkjet technology of titanium oxide nanoparticles deposited by inkjet on fused silica [107]. This approach has led to the concept of interference inkjet printing, providing controlled exotic colored images using a colorless ink prepared from titanium oxide nanoparticles.

As can be seen from the above presented examples, the inkjet printing of nanoparticles has paved a new route for paper electronics, exotic imaging through interference inkjet printing, and for developing photonic nanostructures on flexible substrates. Despite the fact that inkjet printing requires predeposition ink preparation, the technique holds promise as it

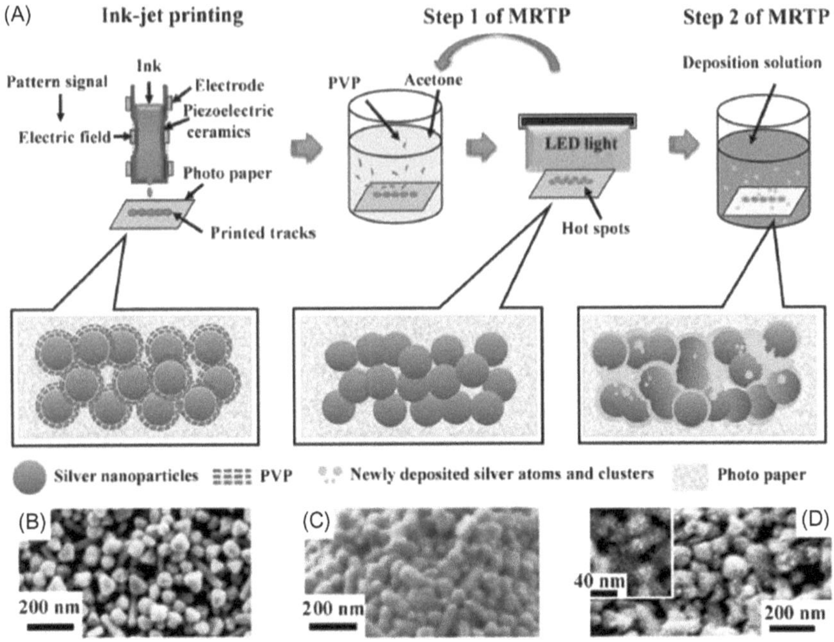

Figure 4.27 (A) Schematic representation of the multistep room temperature post-processing. (B) SEM image of the inkjet deposited silver nanoparticles before sintering, (C) after treatment using step 1 shown in the figure, (D) after treatment with step 2; the insert in (D) shows a high-resolution SEM image of small nanoparticles [106].

is low-cost, compatible with mass production, and hence can be suitable for industrialization.

4.5 Electrospinning for fiber nanofabrication

So far, most of the presented emerging nanofabrication methods have been focusing on semiconductor nanosize materials. However, emerging methods for the fabrication of other classes of nanomaterials are of interest for a variety of industrial application, e.g., fibers, ceramics, polymer, and other composites [108–110].

Electrospinning is a lithography-free desktop fabrication method suitable for the preparation of nanofibers of a variety of materials. Fig. 4.29 above shows the setup of a typical electrospinning apparatus. The electrospinning is suitable for fabricating relatively long one-dimensional nanowires with other techniques. Contrary to other

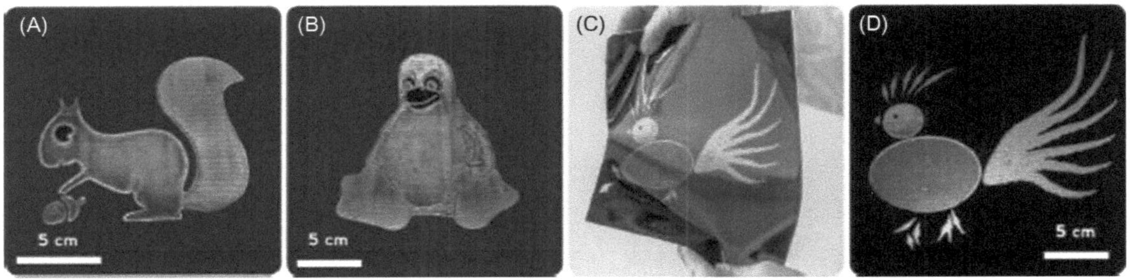

Figure 4.28 Different colored images drawn using the inkjet printing of titanium oxide nanoparticles with the concept of "interference inkjet printing" [107].

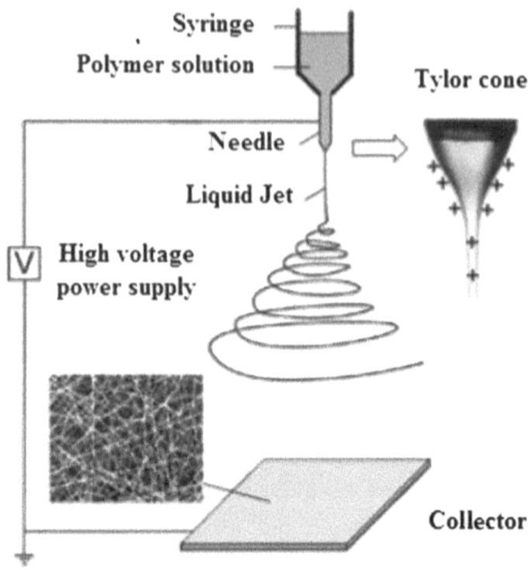

Figure 4.29 Schematic diagram showing a typical setup of electrospinning. The insert to the top right shows the electrified Taylor cone while the insert to the bottom left shows an SEM of a polymer fabricated fiber [108].

one-dimensional nanowires fabrication techniques, the one-dimensional fabrication of nanofibers using electrospinning is relying on the stretching or elongation of a viscous solution. Hence, the diameter of the fabricated nanofiber depends on the properties of the solution and the external applied shear force [108]. It is to be noted that the shear force is exerted due to the repulsion of surface electrostatic charges [108]. Electrospinning is an old technique dating back to 1937, but its use for nanofibers of different materials is rather recent [111].

The electrospinning apparatus is composed of three main basic parts. These are (1) high voltage power supply, either a DC or AC power suppliers can be utilized; (2) a metallic needle, also called the spinneret; and (3) a collector. The metallic spinneret is attached to the solution container, which supplies the required material to fabricate the fiber. The solution can be supplied at a controlled rate, and when a relatively high voltage is applied, the pendent droplet will be electrified with an even distribution of the charge over the nozzle surface [108]. This will lead to a situation where the droplet will be under the action of two major electrostatic forces. The first is Coulomb force due to the external electric field from the applied voltage, and the second is the electrostatic repulsion force caused by the surface charges. Due to these two forces, the pendent solution drop will be distorted into conical shape objects. These conical objects are called Taylor cones [108,112]. At the electric field, when the strength exceeds a threshold value, the electrostatic force will overcome the surface tension of the droplet leading to ejection of the droplet as a jet form the nozzle. This electrified ejection undergoes an elongation and whipping process. As the solution pendent droplets are continuously elongated, the solvent will be evaporated. During this process, the diameter of the electrified liquid jet can be reduced from a few micrometers to tens of a nanometers [108]. This relatively thin ejected jet charged fiber would randomly be deposited as a nonwoven fiber on the collector placed under the spinneret. More than 50 organic polymers have been obtained as a fiber with diameters ranging from a few micrometers down to few tenths of a nanometer [108].

As mentioned above the morphological nature of the fabricated nanofibers depends on two factors; the first is the internal properties of the solution, e.g., elasticity, viscosity, and concentration, and the second is the processing parameters chosen for a specific process, e.g., strength of the electrical field, solution feeding rate, environment temperature, and humidity [108]. To obtain morphologically regular nanofibers the above two mentioned factors have to be optimized carefully for each single process. Otherwise, as commonly observed, beads can appear in different locations along the nanofibers as frequently observed [113]. The beads appear usually due to the effect of three forces during the electrospinning fabrication. The minimization of the surface area leads to the fact that the surface tension will favor the conversion of the liquid jet to a large number of spherical droplets, this is the so-called Rayleigh instability proposed a long time ago [114]. Contrary to this, electrostatic repulsion of similar charges will tend to increase the surface area favoring the formation of beads. The internal viscosity force of the liquid used will resist the formation of beads and favors the formation of a fiber with smooth surface. The balance between these different forces will then determine the formation or disappearance of beads for the fabricated nanofiber. Hence, for the elimination of beads, the surface tension should be suppressed by the effect of repulsion and viscosity forces [108]. Although the primary morphology prepared by electrospinning is thin solid regular wires, other secondary morphologies have also been demonstrated. Hollow, core/sheath porous nanofibers have all been fabricated using the electrospinning approach. Such exotic morphologies have been fabricated due to the invention of new spinnerets and the adjustment of the fabrication parameters.

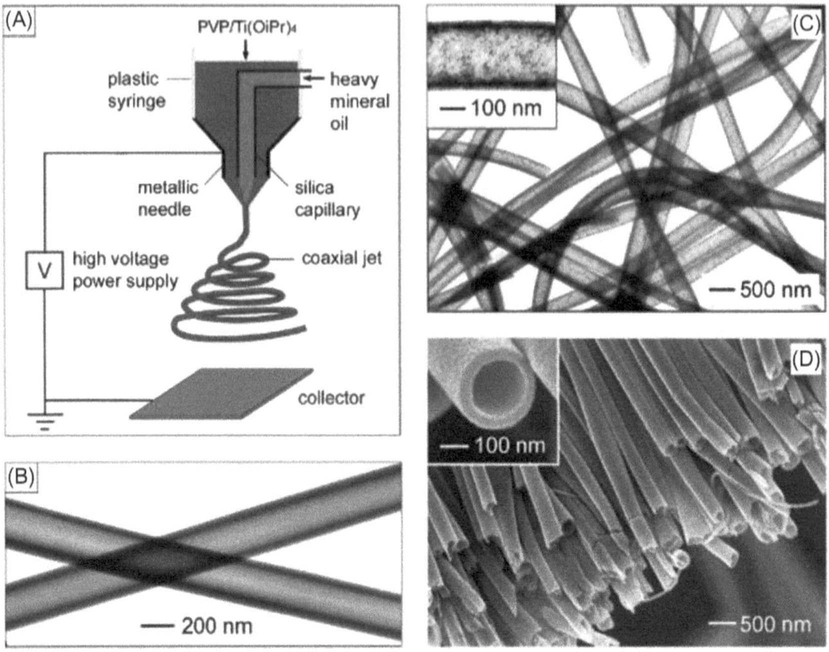

Figure 4.30 (A) Schematic diagram depicting the setup of electrospinning used for the fabrication of hollow nanofibers. As can be seen he spinneret is composed form two coaxial capillaries, through which a heavy mineral oil and ethanol solution containing PVP and $Ti(OiPr)_4$ were ejected simultaneously to form the coaxial jet. (B)–(D) Different TEM and SEM images of the fabricated hollow nanofibers [114]. *TEM*, Transmission elector microscope.

As can be seen in Fig. 4.30, hollow nanofibers with a uniform circular cross section have been demonstrated using the electrospinning method [115]. The hollow morphology was achieved through the injection of two immiscible liquids through coaxial capillary spinneret. Due to the flexibility in fabricating different morphologies, electrospinning has found many applications. It is to be noted that nanofibers fabricated using electrospinning have different features compared to other one-dimensional nanostructures fabricated using other techniques. For example, electrospun nanofibers are highly charged after ejection, and hence can be redirected to be aligned in a required direction. In addition, electrospun nanofibers have extremely long lengths compared to other one-dimensional nanowires, as the ejection process can continue as long as required; nanofibers of several kilometers in length can be fabricated [108].

From the previous presentation, it is clear that electrospinning is a powerful technique for the nanofabrication of fibers with unique properties from a variety of materials. It is also a simple technique regarding instrumentation, is also low-cost, and can be used for mass production. These factors have led to the fact that electrospinning nanofibers have found many applications. To mention some of these applications [108]:

- Smart clothes: due to their better resistance to convective airflow compared to normal clothes, electrospun nanofibers have been demonstrated as candidates for smart clothes [116].
- Supporters for enzymes and catalysts: ceramic and polymeric electrospun nanofibers possess small size and large surface area. Due to these properties, electrospun nanofibers have been successfully utilized as enzyme supporters [117].
- Nanofibers as sensor elements: The possibility of fabricating fluorescent nanofibers have successfully led to the development of optical sensors [118].
- Biomedical applications: electrospun nanofibers scaffolds have been attractive for many biomedical applications due to their unique geometrical arrangements [110].

From the above presentation, it is clear that electrospinning as a technique for the fabrication of nanofibers can be utilized for many applications. This is due to the unique features of the fabricated nanofibers, the wide possible material classes that can be used, and the fact that it is a scalable process with low cost. Nevertheless, before electrospinning can be a standard industrial approach, different issues are to be solved. Among these issues are the improved control of the size and morphology of the produced nanofibers. Another critical issue is the diameter of the produced fiber, at the moment it is not straightforward to fabricate fibers with a diameter in the deep sub-100 nm. To achieve fibers with diameters in the deep sub-100 nm, improvement of the instrumentation as well as theoretical modeling is highly recommended [108].

4.6 Chemical nanofabrication methods

Chemical routes for the fabrication of nanostructures is emerging as a popular choice because most of them are scalable and in addition, they constitute a low cost approach. Chemical methods are also divided into top-down and bottom-up categories. Below we present and discuss the most common of chemical routes of the two categories.

4.6.1 Chemical top-down nanofabrication methods

During the past few decades, and as discussed in Chapter 3, Conventional nanofabrication methods, the use of top-down physical methods for the synthesis of nanostructures has been the dominating route. Mainly it was and still the adopted industrial method for the fabrication of electronic components. On the other hand, the chemical top-down route has not been heavily explored [119]. As most of the physical top-down approaches require extremely controlled laboratory environments, i.e., clean room laboratories, which in turn imply high cost, researchers have been intensively working to develop affordable alternative routes. Chemical approaches has been the obvious choice.

Nevertheless, and in general, the utilization of chemistry as a route to fabricate nanostructures have tremendously exploded during the past few years. This can be seen from the relatively large number of published papers on the topic of chemical synthesis of nanostructures. Due to the easiness and low cost, both top-down and bottom-up chemical routes have been increasingly of interest to adopt and develop new methods that can be utilized for the fabrication of nanostructures in the deep sub-100 nm. Below we will present and discuss the most popular top-down chemical approaches utilized recently for the fabrication of nanostructures. Fig. 4.31 shows the main top-down chemical nanofabrication methods [119]. As we it was presented in this chapter and Chapter 3, Conventional nanofabrication methods, top-down physical approaches

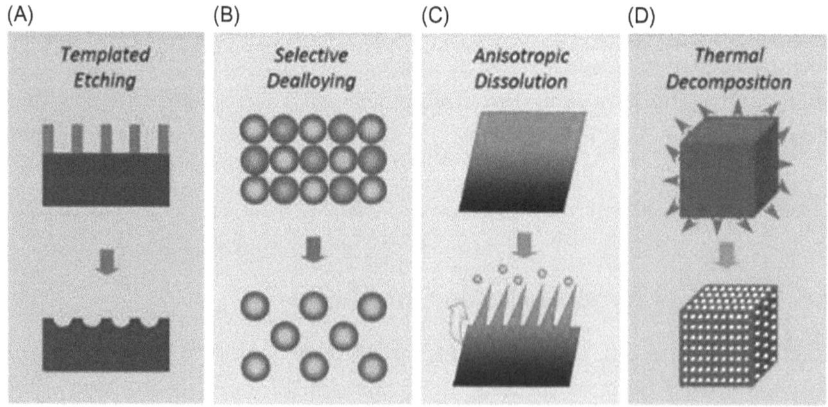

Figure 4.31 Chemical routes to top-down nanofabrication: (A) template etching, (B) selective de-alloying, (C) anisotropic dissolution, and (D) thermal decomposition [119].

have demonstrated great success in demonstrating precise controlled nanostructures for a variety of applications. However, most o0f these techniques required controlled laboratory environments, and in addition, other processing conditions are costly like high vacuum, extreme high temperatures etc. to the contrary and as will be presented below, top-down chemical methods can be used to fabricate small size structures from solids, and liquids in much less restricted processing conditions [119]. The top-down chemical approaches to be presented here are limited to four cases. These methods can generate porous materials, and other order three-dimensional nanostructures arrays [119]. In particular four different approaches are to be discussed and presented. These are template etching, selective de-alloying, anisotropic dissolution, and thermal decomposition.

In template etching nanostructures can be fabricated by using a "template" as a mask for chemical etching. The template is to direct the chemical etching. After the etching step, the nanoscale pattern will appear after the removal of the template [119]. This approach has been successful in creating complex three-dimensional arrays on solid substrates.

Fig. 4.32 above show a schematic diagram of the steps of processing such arrays in 3 dimensions on the surface of a solid substrate [120]. While in Fig. 3.34, a typical example of three-dimensional array of

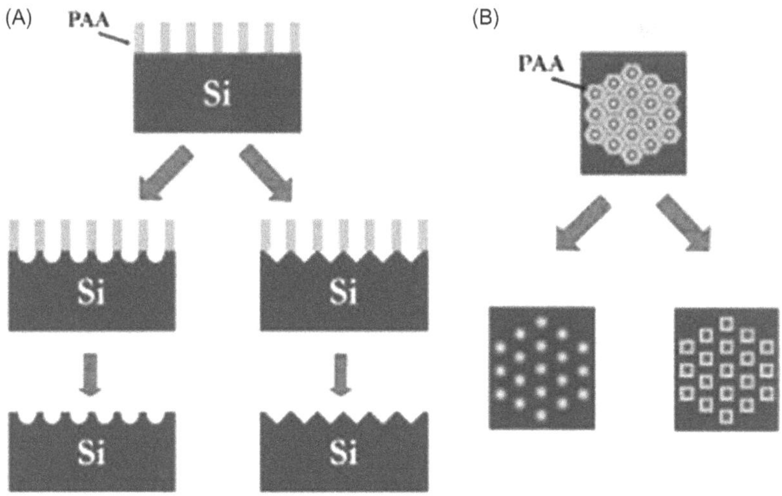

Figure 4.32 (A) Top view schematic representation of the main process steps for Si nanopatterning through a porous anodic alumina-masking layer. (B) Top view of (A) [120].

Figure 4.33 SEM images at two different magnifications of hexagonally arranged arrays of concave pits on a preselected area on Si, fabricated by electrochemical etching in IPA:HF:H2O 10:3:6 solution at 15 mA/cm^2 for 90 s through a porous anodic aluminum mask [120].

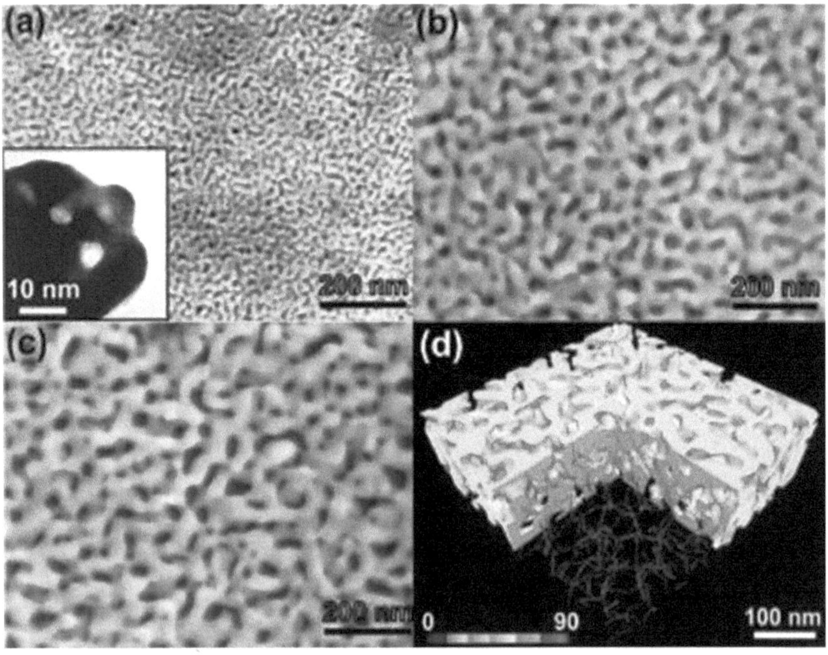

Figure 4.34 SEM of nanoporous gold with ore sizes of (A) 14 nm, (B) 35 nm, and (C) 47 nm. The insert in (A) shows a TEM image of a nanoporous section of one sample. (D) Three-dimensional sketch of the nanoporous gold [121]. *TEM*, Transmission elector microscope.

nanostructures fabricated using the template method is presented [120]. In this example porous anodic aluminum (PAA) is prepared by anodization of aluminum and is used to fabricate highly ordered hexagonal three-dimensional array of nanostructures as shown in the SEM of Fig. 4.33 [120].

The second chemical top-down approach to be presented is the selective de-alloying. In selective de-alloying, also called demetallification, selective removal of one metal from a metal alloy is performed. The selective de-alloying will obviously lead to porous structures. As it is known, porous structures are structures with extremely small pores and possesses relatively high surface area to volume ratio and porous structures are quite useful and efficient for many applications like e.g., sensing. Fig. 4.34 shows an example of nanoporous structure demonstrated using the selective de-alloying [121]. The nanoporous shown in this figure were achieved by chemically de-alloying 100 nm thick $Au_{35}Ag_{65}$ gold leaf using 70% volume HNO_3 for different durations of time at 25°C [121]. As can be seen the selective de-alloying can produce porous structures with dep sub-100 nm pores, however, the disadvantages is that it is limited to alloys.

The third chemical top-down approach to be resented the anisotropic dissociation approach. The fabrication of, or more accurately synthesis of, nanocrystals or generally nanostructures in solutions via nucleation has been the most investigated chemical nanofabrication method and it will be discussed in the next chapter due to its potential in the development of nanoscience in general. The term chemical synthesis of nanostructures is in fact the bottom-up chemical synthesis approach. In a reverse process, relatively large crystals are dissolved in a very controlled process in undersaturated solutions to fabricate nanoscale small crystals. Fig. 4.35 illustrates the process of fabrication of nanosized array via the anisotropic dissociation approach. This is a new approach; however, the utilization of it is limited to few materials.

The final top-down chemical approach to be presented is the thermal decomposition. In thermal decomposition hear is used to dissociate chemical bonds of a compound material. This process can lead to the fabrication of three-dimensional porous nanomaterial with controlled pores. In this respect, it is a unique being a "one step" process for providing controlled porous material [119].

High-quality cadmium carbonate microcrystals were transformed into high purity cadmium oxide porous structure. This was achieved by

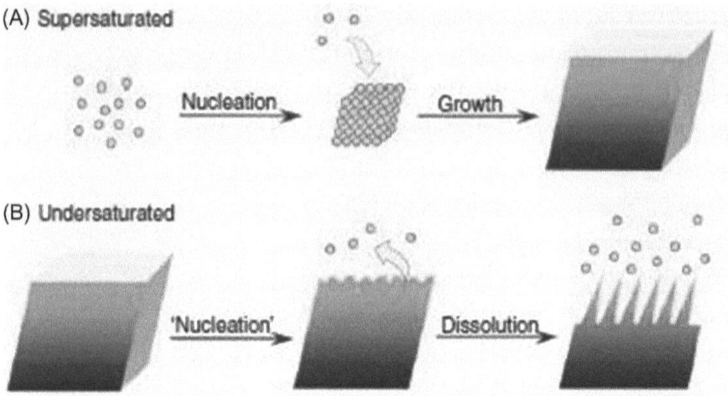

Figure 4.35 Schematic representation of (A) crystal growth, and (B) dissociation in supersaturated and unsaturated solutions. Respectively [122].

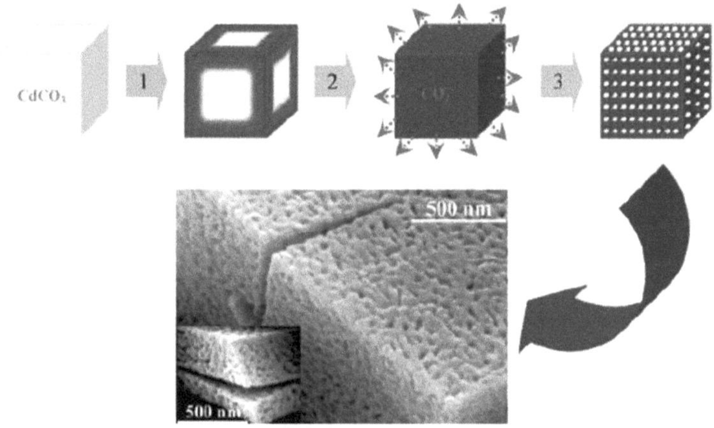

Figure 4.36 Schematic representation of the formation of nanoporous cadmium oxide by a one-step decomposition of cadmium carbonate microcrystals. Cadmium oxide is fir formed at the corners edges. Then the formation of the cadmium oxide process form the outer surface to the inner part of the crystal. Eventually a porous cadmium oxide architecture is formed. The SEM shows the final structure of the porous nanomaterial [123].

annealing at 500°C for 30 minutes [123]. As can be seen in Fig. 4.36, the final fabricated structures is a porous structure with small pores. The disadvantage of this method is that it is restricted to compound materials. Also the starting material has to be of very high quality in order to achieve porous structure [123].

4.6.2 Chemical bottom-up nanofabrication methods

The nucleation of nanocrystals from solutions at low temperature, i.e., bottom-up chemical nanofabrication is the most intensively investigated method in recent years. This is due to the expected potential in proving alternative low cost approach to different application in nanoscience. This is why a variety of different low-temperature chemical methods have been developed and tested. Due to this, Chapter 5 will be devoted for low-temperature chemical nanofabrication methods.

4.7 Preserving nanostructures

In many processes, nanomaterials are fabricated in the presence of water. Water molecule is approximately 0.35 nm (3.5 Å) in size [124]. Hence, a single nanotube of 50 nm length is equivalent about 150 water molecules in length, and in many experiments menisci of water were observed. The presence of water inside nanomaterials can lead to water–water and water–solid interactions [125]. It is often attributed that these water menisci is responsible for many effects, e.g., causing capillary pressure and drying stresses in nanoporous solids. Hence, developing drying methods for nanomaterials is important. The readers can find reviews of modern drying methods in [124,126,127].

References

[1] G. Bachmann, Interdisciplinary co-operation: a critical need in nanotechnology, Asia Pacific Tech Monitor 19 (5) (2002) 34–40.
[2] R. Comparo, Nanotechnology: what is at stake, Asia Pacific Tech Monitor 5 (19) (2002) 25–33.
[3] G. Seniutinas, A. Balčytis, I. Reklaitis, F. Chen, J. Davis, C. David, et al., Nanophotonics 6 (2017) 923.
[4] J. Fischer, M. Wegener, Laser Photon. Rev. 7 (2013) 22.
[5] Z. Gan, Y. Cao, R.A. Evans, M. Gu, Three-dimensional deep sub-diffraction optical beam lithography with 9 nm feature size, Nat. Commun. 4 (2013). Article number: 2061, 1–7.
[6] L. Rapp, B. Habrl, C.J. Pickard, J.E. Bradby, E.G. Gamaly, J.S. Williams, et al., Nat. Commun. 6 (2015). Article number: 7555.
[7] A. Vailionis, E.G. Gamaly, V. Mizeikis, W. Yang, A.V. Rode, S. Juodkazis, Nat. Commun. 2 (2011) 445.
[8] Available from: https://www.edn.com/electronics-blogs/edn-moments/4402146/1st-text-message-is-sent--December-3--1992.
[9] W. Bru, E.-B. Kley, B. Schnabel, I. Stolberg, M. Zierbrock, R. Plontke, Microelectron. Eng. 27 (1995) 135.
[10] R. Bonam, P. Verhagen, A. Munder, J. Hartley, J. Vac. Sci. Technol. 28 (2010) 34.
[11] J. Kretza, L. Dreeskornfelda, Microelectron. Eng. 61 (2002) 607.

[12] Y. Mitsuia, T. Sunaoshia, J.C. Leeb, Microelectron. Reliab. 49 (2009) 1182.
[13] D.J. Thompson, D. Murphy, R.W. Speirs, R.M.W. van Bijnen, A.J. McCulloch, R. E. Scholten, et al., Phys. Rev. Lett. 117 (2016) 193202.
[14] A.J. McCulloch, B.M. Sparkes, R.E. Scholten, J. Phys. B. 49 (2016) 164004.
[15] J. Vila-Comamala, S. Gorelick, V.A. Guzenko, C. David, J. Vac. Sci. Technol. 29 (2011) 06F301.
[16] V.R. Manfrinato, L. Zhang, D. Su, H. Duan, R.G. Hobbs, E.A. Stach, et al., Nano Lett. 13 (2013) 1555.
[17] B. Cord, J. Yang, H. Duan, D. Joy, J. Klingfus, K. Berggren, J. Vac. Sci. Technol. B 27 (2009) 2616.
[18] Y. Liao, Practical Electron Microscopy and Database, an Online Book, 2007.
[19] R. Howard, H. Craighead, L. Jackel, P. Mankiewich, M. Feldman, J., Vac. Sci. Technol. B 1 (1983) 1101.
[20] M. Isaacson, A. Murray, J., Vac. Sci. Technol. 19 (1981) 1117.
[21] C. Vieu, F. Carcenac, A. Pepin, Y. Chen, M. Mejias, A. Lebib, et al., Appl. Surf. Sci. 164 (2000) 111.
[22] W.F. van Dorp, B. van Someren, C.W. Hagen, P. Kruit, Nano Lett. 5 (2005) 1303.
[23] H. Duan, V. Manfrinato, J. Yang, D. Winston, B. Cord, K.J. Berggren, Vac. Sci. Technol., B 28 (2010) C6H11.
[24] G. Hughes, L.C. Litt, A. Wuest, S. Palaiyanur, Proc. SPIE 7028 (2008).
[25] T.R. Groves, R.A. Kendall, J. Vac. Sci. Technol. B 16 (1998) 3168.
[26] M. Muraki, S. Gotoh, J. Vac. Sci. Technol. B 18 (2000) 3061.
[27] D. Chao, A. Patel, T. Barwicz, H.I. Smith, R. Menon, J. Vac. Sci. Technol. B 23 (2005) 2657–2661.
[28] T.L. Andrew, H.-Y. Tsai, R. Menon, Science 324 (2009) 917.
[29] T.F. Scott, B.A. Kowalski, A.C. Sullivan, C.N. Bowman, R.R. McLeod, Science 324 (2009) 913.
[30] L. Li, R.R. Gattass, E. Gershgoren, H. Hwang, J.T. Fourkas, Science 324 (2009) 910–913.
[31] N. Fang, H. Lee, C. Sun, X. Zhang, Science 308 (2005) 534.
[32] P. Srisungsitthisunti, O.K. Ersoy, X.F. Xu, Appl. Phys. Lett. 98 (2011) 223106.
[33] Z. Liu, H. Lee, Y. Xiong, C. Sun, X. Zhang, Science 315 (2007) 1686.
[34] W. Srituravanich, L. Pan, Y. Wang, C. Sun, D.B. Bogy, X. Zhang, Nat. Nanotechnol. 3 (2008) 733.
[35] L. Pan, Y. Park, Y. Xiong, E. Ulin-Avila, Y. Wang, L. Zeng, et al., Sci. Rep. 1 (2011) 132.
[36] N.B. Nguyen, M.P. Nielsen, L. Lafone, E. Clarke, P. Fry, R.F. Oulton, Appl. Phys. Lett. 111 (2017) 261107.
[37] E. Nicoletti, D. Bulla, B. Luther-Davies, M. Gu, Nano Lett. 11 (2011) 4218–4221.
[38] P. Zijlstra, J.W.M. Chon, M. Gu, Nature 459 (2009) 410–413.
[39] S. Kawata, H.-B. Sun, T. Tanaka, K. Takada, Nature 412 (2001) 697–698.
[40] Y. Cao, Z. Gan, B. Jia, R.A. Evans, M. Gu, Opt. Exp. 19 (2011) 19486–19494.
[41] M. Malinauskas, M. Farsai, A. Piskarskas, S. Juodkazis, Phys. Rep. 533 (2013) 1.
[42] E. Altewischer, M.P. van Exter, J.P. Woerdman, Nature (London) 418 (2002) 304.
[43] A.N. Grigorenko, A.K. Geim, Gleeson, Y. Zhang, H.F. Firsov, A.A. Khrushchev, J. Petrovic, Nature 438 (2005) 335.
[44] W. Han, L. Jiang, X. Li, Q. Wang, S. Wang, J. Hu, et al., Sci. Rep. 7 (2017) 17333.
[45] S.V. Makarov, V.A. Milichko, I.S. Mukhin, I.I. Shishkin, D.A. Zuev, A.M. Mozharov, et al., Laser Photon. Rev. 10 (2016) 91–99.
[46] R.E.J. Watkins, P. Rockett, S. Thoms, R. Clampitt, R. Syms, Vacuum 36 (1986) 961.
[47] C.-S. Kim, S.-H. Ahn, D.-Y. Jang, Vacuum 86 (2012) 1014.

[48] M. Albrecht, C.T. Rettner, A. Moser, B.D. Terris, Microsyst. Technol. 13 (2007) 129.

[49] R. Mühle, Rev. Sci. Instrum. 63 (1992) 3040.

[50] A. Taurino, M. Catalano, M. Lomascolo, A. Persano, A. Convertino, L. Cerri, Phys. E 41 (2009) 734.

[51] J. Yu, J. Liu, J. Zhang, J. Wu, Mater. Lett. 206 (2006) 60.

[52] A. Kempe, R. Wirth, W. Altermann, R.W. Stark, J.W. Schopf, W.M. Heckl, Precambrain Res. 140 (2005) 36.

[53] J. Melngailis, Critical review, J. Vac. Sci. Technol. B 5 (1987) 469.

[54] A.D. Dubner, A. Wagner, J. Melngailis, C.V. Thompson, J. Appl. Phys. 70 (1991) 665.

[55] Available from: www.srim.org.

[56] I. Utke, P. Hoffmann, J. Melngailis, J. Vac. Sci. Technol. B 26 (2008) 1197.

[57] M.V. Gorkunov, A.A. Ezhov, V.V. Artemov, O.Y. Rogov, S.G. Yudin, Appl. Phys. Lett. 104 (2014) 221102.

[58] A.V. Kondratov, O.Y. Rogov, R.V. Gainutdinov, Ultramicroscopy 181 (2017) 181.

[59] O.Y. Rogov, V.V. Artemov, M.V. Gorkunov, A.A. Ezhov, D.N. Khmelenin, J. Microscopy 268 (2017) 254.

[60] H.-P. Phan, T. Kozeki, T. Dinh, T. Fujii, A. Qamar, Y. Zhu, et al., RSC Adv. 5 (2015) 82121.

[61] Ch Akhmadaliev, L. Bischoff, B. Schmidt, Mater. Sci. Eng. C 26 (2006) 818.

[62] M. Kiuchi, S. Matsui, Y.J. Isono, Microelectromech. Syst. 16 (2007) 191.

[63] J.F. Einsle, J.-S. Bouillard, W. Dickson, A.V. Zayats, Nanoscale Res. Lett. 6 (2011) 572.

[64] X. Chen, C.-B. Chen, C.N.B. Udalagama, M. Ren, K.E. Fong, L.Y.L. Yung, et al., Biophys. J. 104 (2013) 1419.

[65] S.A. Boden, T.M.W. Franklin, L. Scipioni, D.M. Bagnall, H.N. Rutt, Microsc. Microanal. 18 (2012) 1253.

[66] Z. Mi, Y. Zhang, S.K. Vanga, C.-B. Chen, H.Q. Tan, F. Watt, et al., Nat. Commun. 6 (2015) 8832.

[67] G. Gervinskas, G. Seniutinas, S. Juodkazis, Laser Photon. Rev. 7 (2013) 1049.

[68] A. Dazzi, C.B. Prater, Q. Hu, D.B. Chase, J.F. Rabolt, C. Marcott, Appl. Spectrosc. 66 (2012) 1365.

[69] R. Hillenbrand, T. Taubner, F. Keilmann, Nature 159 (2002) 418.

[70] G. Binnig, H. Rohrer, IBM J. Res. Dev. 30 (1986) 355.

[71] Available from: https://www.nobelprize.org/nobel_prizes/physics/laureates/1986/press.html.

[72] J. Guo, K. Bian, Z. Lin, Y. Jiang, J. Chem. Phys. 145 (2016) 160901.

[73] G. Gramse, A. Kölker, T. Lim, T.J.Z. Stock, H. Solanki, S.R. Schofield, et al., Sci. Adv. 3 (2017) e1602586.

[74] H.M. Pollock, A. Hammiche, J. Phys. D: Appl. Phys. 34 (2001) R23.

[75] S. Sadat, A. Tan, Y.J. Chua, Nano Lett. 10 (2010) 2613.

[76] G. Baffou, M.P. Kreuzer, F. Kulzer, R. Quidant, Opt. Exp. 17 (2009) 3291.

[77] P. Löw, B. Kim, N. Takama, C. Bergaud, Small 4 (2008) 908.

[78] F. Vetrone, R. Naccache, A. Zamarrón, A. Juarranz de la Fuente, F. Sanz-Rodríguez, L. Martinez Maestro, et al., ACS Nano 4 (2010) 3254.

[79] G. Baffou, P. Bon, J. Savatier, J. Polleux, M. Zhu, M. Merlin, et al., Nano Lett. 6 (2012) 2452.

[80] I.A. Vartanyants, J.A. Pitney, J.L. Libbert, I.K. Robinson, Reconstruction of surface morphology from coherent x-ray reflectivity, Phys. Rev. B 55 (1997) 13193–13202.

[81] M. Sutton, S.G.J. Mochrie, T. Greytak, S.E. Nagler, L.E. Berman, G.A. Held, et al., Nature 352 (1991) 608.

[82] P. Thibault, M. Dierolf, A. Menzel, O. Bunk, Ch David, F. Pfeiffer, Science 321 (2008) 379.
[83] C. Stephenson, A. Hubler, Sci. Rep. 5 (2015) 15044.
[84] G.M. Whitesides, B. Gryzbowski, Science 295 (2002) 2418.
[85] Z. Nie, A. Petukhova, E. Kumacheva, Nat. Nanotechnol. 5 (2010) 15.
[86] K. Thorkelsson, P. Bai, T. Xu, Nano Today 10 (2015) 48.
[87] A.S. Iyer, K. Paul, Nanobiotechnology 9 (2015) 122.
[88] C.T. O'Mahony, R.A. Farell, J.D. Holmes, M.A. Morris, The thermodynamics of defect formation in self-assembled systems, Chapter 13 (Open access), in: Thermodynamics—Systems in Equilibrium and Non-Equilibrium, pp. 279–306. doi:10.5772/20145. ISBN 978-953-307-283-8.
[89] K. Ariga, J.P. Hill, M.V. Lee, A. Vinu, R. Charvet, S. Acharya, Sci. Technol. Adv. Mater. 9 (2008) 014109.
[90] A.M. Sweetman, S.P. Jarvis, H. Sang, I. Lekkas, P. Rahe, Y. Wang, et al., Nat. Commun. 5 (2013) 1–7.
[91] Z.H. Nie, D. Fava, E. Kumacheva, S. Zou, G.C. Walker, M. Rubinstein, Nat. Mater. 6 (2007) 609.
[92] M. Rycenga, J.M. McLellan, Y. Xia, Adv. Mater. 20 (2008) 2416.
[93] M.A. Correa-Duarte, J. Perez-Juste, A. Sanchez-Iglesias, M. Giersig, L.M. Liz-Marzan, Angew. Chem. Int. Ed. 44 (2005) 4375.
[94] J. Arlt, V.A. Martinez, A. Dawson, T. Pilizota, W.C.K. Poon, Nat. Commun. 9 (2018) 768.
[95] J.C. Love, L.A. Estroff, J.K. Kriebel, R.G. Nuzzo, G.M. Whitesides, Chem. Rev. 105 (2005) 1103.
[96] Y. Liu, Nat. Chem. 9 (2017) 1037.
[97] F. Zhang, J. Nangreave, Y. Liu, H. Yan, J. Am. Chem. Soc. 136 (2014) 11198.
[98] N.C. Seeman, J., Theor. Biol. 99 (1982) 237.
[99] C.R. Simmons, F. Zhang, T. MacCulloch, N. Fahmi, N. Stephanopoulos, Y. Liu, et al., J. Am. Chem. Soc. 139 (2017) 11254.
[100] S. Alan, H. Jiang, H. Shahbazbegian, J.N. Patel, B. Kaminska, Nanotechnology 28 (2017) 135302.
[101] J.W. Strutt, Proc. London Math. Soc 10 (1878) 4.
[102] E. Sowade, E. Ramon, K.Y. Mitra, C. Martínez-Domingo, M. Pedró, J. Pallarès, et al., Sci. Rep. 6 (2016) 1.
[103] A.D. Furasova, V. Ivanovski, A.V. Yakovlev, V.A. Milichko, V.V. Vinogradov, A. V. Vinogradov, Nanoscale 9 (2017) 13069.
[104] C. Novara, F. Petracca, A. Virga, P. Rivolo, S. Ferrero, A. Chiolerio, et al., Nanoscale Res. Lett. 9 (2014) 527.
[105] S.R. Das, Q. Nian, A.A. Cargill, J.A. Hondred, S. Ding, M. Saei, et al., Nanoscale 8 (2016) 15870.
[106] X.-Y. Zhang, J.-J. Xu, J.-Y. Wu, F. Shan, X.-D. Ma, Y.-Z. Chen, et al., RSC Adv. 7 (2017) 8.
[107] A.V. Yakovlev, V.A. Milichko, Vinogradov, A.V. Vinogradov, ACS Nano 10 (2016) 3078.
[108] D. Li, Y. Xia, Adv. Mater. 16 (2004) 1151.
[109] R. Jayakumar, M. Prabaharan, S.V. Nair, M. Tamura, Biotechnol. Adv. 28 (2010) 142.
[110] D. Liang, B.S. Hsiao, B. Chu, Adv. Drug Deliev. Rev. 59 (2007) 1392.
[111] A. Formalas, US Patent 1 975 504, 1934.
[112] D.H. Reneker, I. Chun, Nanotechnology 7 (1996) 216.
[113] H. Fong, I. Chun, D.H. Reneker, Polymer 40 (1999) 4585.
[114] L. Rayleigh, Proc. R. Soc. Lond. 14 (1882) 184.

[115] D. Li, Y. Xia, Nano Lett. 4 (2004) 933.
[116] P. Gibson, H. Schreuder-Gibson, D. Rivin, Colloids Surf. A 187 (2001) 469.
[117] H. Jia, G. Zhu, B. Vugrinovich, W. Kataphinan, D.H. Reneker, P. Wang, Biotechnol. Prog. 18 (2002) 1027.
[118] H. Liu, J. Kameoka, D.A. Czaplewski, H.G. Craighead, Nano Lett. 4 (2004) 671.
[119] H.-D. Yu, D. Regulacio, E. Ye, M.-Y. Han, Chem. Sov. Rev. 42 (2013) 6006.
[120] F. Zacharatos, V. Gianneta, A.G. Nassiopoulou, Nanotechnology 19 (2008) 495306.
[121] T. Fujita, H. Okada, K. Koyama, K. Watanabe, S. Maekawa, M.W. Chen, Phys. Rev. Lett. 101 (2008) 166601.
[122] H.D. Yu, D. Yang, D. Wang, M.Y. Han, Adv. Mater. 22 (2010) 3181.
[123] H.D. Yu, D. Wang, M.Y. Han, J. Am. Chem. Soc. 129 (2007) 2333.
[124] Z. Pakowski, Transp. Porous Med. 66 (2007) 19.
[125] Y. Gogotsi, J.A. Libera, A.G. Yazicioglu, C.M. Megaridis, Appl. Phys. Lett. 79 (2001) 1021.
[126] B. Wang, W. Zhang, W. Zhang, A.S. Mujumdar, L. Huang, Drying Technol. 37 (2005) 7.
[127] J. Beirowski, S. Inghelbrecht, A. Arien, H. Gieseller, J. Pharm. Sci. 5 (2011) 1958.

Further reading

L. Jacot-Descombes, M.R. Gullo, V.J. Cadarso, J. Brugger, J. Micromech. Microeng. 22 (2012) 074012.

CHAPTER 5

Low-temperature chemical nanofabrication methods

5.1 Introduction

In the preceding chapters, the most common conventional and emerging top-down techniques for processing deep sub-100 nm features have been presented and discussed. Low-temperature chemical-based bottom-up approaches, which are among the most researched topics now have been omitted due to its importance owing to the ongoing huge interest and will be the topic of this separate chapter. The conventional and emerging top-down approaches presented and discussed in the previous chapter(s), although providing an acceptable route for achieving deep sub-100 nm feature sizes, have some disadvantages. The main disadvantage is that they are mostly relying on a complicated costly lithographic approach. In addition, they are serial approaches and cannot be utilized for mass production. Hence, they can only be suitable for research and not for industrial production. This implies that to adopt top-down micro/nanolithography techniques and other vapor deposition tools for developing nanostructures and nanodevices with feature sizes below 100 nm will not be a feasible route. On the other hand, such feature sizes are quite normal for biologists. In biological sciences, such small feature size prevails in quite complex configuration in biological systems mostly through self-assembly. Examples of these deep sub-100 nm entities are viruses. On the other hand, chemists deal with configurations of attached molecules and ions that can range from a single molecule/ion up to few hundreds/thousand molecules/ions, also attached through self-assembly. Hence a feasible route for the development and utilization of deep sub-100 nm feature objects and devices could prevail by the use of chemistry following the biological self-assembly route but by utilizing reactions that do not rely on biological catalysts or coded genes [1].

It is obvious that, the real dream of any nano-scientist is to develop a "well" controlled required "large" size object starting from self-assembling single molecules. Here by "large" size objects it is meant objects with

Low Temperature Chemical Nanofabrication
DOI: https://doi.org/10.1016/B978-0-12-813345-3.00005-8

features of many micrometers or much larger. Such large size objects utilized by molecular manufacturing (MM) are not yet achievable through molecular chemical synthesis and can only be seen in "science fiction" films. While objects with lower sizes, that is, in the range of a few hundreds of nanometers down to a few nanometers are continuously being utilized by the use of low-temperature chemical reactions. Now chemists are very active in achieving synthesis recipes and reactions to assemble molecules in a well-controlled configuration based on noncovalent bonding. The choice of noncovalent bonding here is because most of the nanomaterials of interest fall in the class of inorganic materials, being of a semiconducting, conducting, or insulating nature. The present worldwide activities of chemists in developing mature recipes for the synthesis of nanostructures will no doubt widen the scope of chemistry to include it as a tool for extending chemistry for solving different unsolved issues in materials and biological sciences [1].

Having an inorganic nanomaterial through a chemical route implies the design of a short-range well-controlled reaction. In conventional inorganic chemistry, usually reactions and recipes are optimized to overcome barriers that hinder the required reaction or process. These barriers are all connected to the energy involved in the required reaction. Examples of these barriers are the thermal heat needed, the inherently large diffusion path lengths, etc. Such a strategy of optimization, although it works well for bulk materials, cannot be utilized for the chemical synthesis of nanostructures [2]. The main reason for this is that the synthesis of nanostructures requires recipes that lead to three-dimensional crystals of relatively small size at a much lower temperature and milder conditions. To utilize chemistry for the synthesis of nanostructures, recipes and procedures based on molecular precursors have been developed. Such choices have led to reduced diffusion-lengths and usually yield well-defined three-dimensional nanomaterials under milder conditions.

By transferring a short-range chemical order present in precursors to three-dimensional infinite length many successful recipes for the synthesis of nanostructures through molecular low-temperature chemical routes have been demonstrated [3–5]. Nevertheless, it is important to mention that the kinetics of the synthesis process are not well understood in most of the demonstrated recipes. Such lack of knowledge will restrict the predictability of the outcome for the situations where an inorganic nanomaterial is to be produced. This is in contrast to organic materials [3]. Despite this low knowledge of the kinetics of the processes for inorganic nanomaterials synthesis, chemists have succeeded in designing successful recipes and have demonstrated a variety of stable relatively well-controlled inorganic nanomaterials by the rational design of the different synthesis parameters [4–6]. The preceding discussion on the synthesis of inorganic materials can be illustrated in two cases, as shown in Fig. 5.1A and B. In the first case, that is, Fig. 5.1A, the outcome (OUT 1–3) are not controlled. Here the synthesis parameters will lead to production of a chemical product that is accessible to the thermodynamic situation available. This will lead to a chemical product with abnormal growth, and phase segregation, mixed elements, formation of side-products, etc. will all prevail. On the other hand, when molecular precursors' seeds are used more control on the product and its phase is achieved. Here by controlling the synthesis parameters, more control of the required final material and at the same time reduction of unwanted reaction paths will be easy to demonstrate. This case is demonstrated in Fig. 5.1B, where many chemical control valves are utilized. The use of chemistry to successfully synthesize nanomaterials at low temperature, that is, at temperatures at around 100°C or lower, is mainly due to the use of molecular precursors [2]. The main rea son for this is that molecular precursors are formed of elements that are chemically linked, and hence they precipitate as solids without any need for diffusion. Adding to this is the fact that the collision path length for such molecular precursors is relatively short compared to that of conventional chemical procedures. This is the basic main argument for the advantages of the use of the chemical approach for nanomaterial synthesis utilizing molecular precursors. Hence, a solid-state nanostructure can be obtained by a careful choice of a molecular precursor containing the proper ratio for the required composition having the required elements of the targeted solid-state nanostructure [7]. In this case, and as mentioned above, the nucleation barrier of the synthesis will be at a minimum and the growth will be facilitated. Although this might sound simple, the real

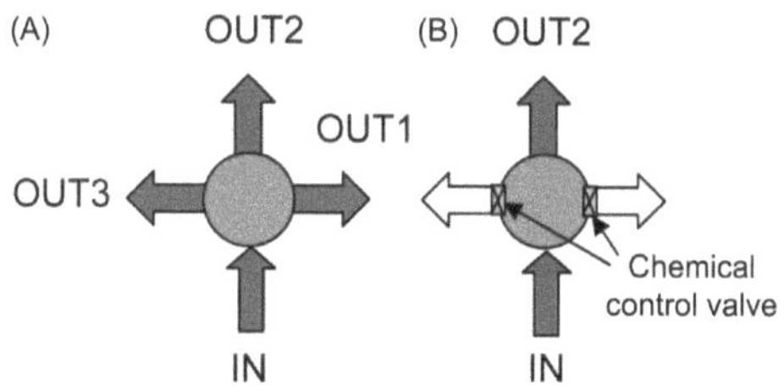

Figure 5.1 Outcome control in (A) conventional and (B) molecular level synthesis [2].

situation depends on many other factor concerning the synthesis environment, as will be discussed below. In this way, the use of chemical routes at low temperature, that is, soft chemistry, can lead to the selective synthesis of the required material in its metastable state. Further, by adjusting the synthesis environment, soft chemistry can also allow the tuning of the nanostructure morphology [7].

The most important issues to be considered when synthesizing nanostructures is the control of the size, morphology, spatial distribution, purity, and reproducibility. Researchers have in recent years developed different methods and varied different synthesis parameters to achieve control over size, morphology distribution, etc. Indeed, many different nanostructures have been demonstrated and reproduced by others; however, still much more research work is needed to have well-documented information regarding the kinetics of the synthesis processes. In recent years many different methods and approaches using molecular precursors have been developed. Below we will present the most common chemical low-temperature methods that have been of interest recently.

5.2 The chemical precipitation method

The use of chemical precipitation as a route to achieve well-controlled solids in a particle form has been as old as colloidal science. In general, chemical precipitation is mainly utilized to synthesize monodispersed particles of different morphology, but usually spherical shape is the most common. Monodispersed particles have existed in nature around us since the creation of mankind. Opals with their amazing iridescence is one example [8]. This amazing iridescence is due to the opals' structure since they are composed of an aggregate of silica particles. All naturally occurring colloidal nanoparticles (NPs) through chemical precipitation have existed due to the fact that the conditions for their formation was satisfied. Chemists have for a long time, and to be more specific since more than 100 years ago, produced colloidal dispersed monoparticles of different sizes using the chemical precipitation method. Before the discovery of the modern submicrometer spectroscopy direct observation instruments, for example, scanning electron microscopy, these old-days colloidal monodispersed particles where characterized by indirect methods, like the Tyndall effect, etc. [8]. Some are of an elemental nature like gold, selenium, sulfur, etc. and some are inorganic compounds like barium sulfate, sliver halides, calcium fluoride, etc. [8]. Most of the previous research work was on hydrous metal oxides as they are of interest for many applications like catalysis, filters, pigments, and coatings. Although intensive research was carried out previously, the formation processes studies have yielded elusive information. As these processes were carried out in a liquid media, and hence complexity arises due to many factors. This elusive information is due to the complexity when one deals with sols. Sol refers to a solution containing homogenously dispersed particles, mainly of spherical shape. This is because the chemical behavior of sols is very sensitive to many parameters. Obviously, the composition, and concentration, of solutes are sensitive to many factors, as temperature, pH, and the concentration of anionic species are some of the factors that play an important role and dictate the final product. Hence, for a homogenous precipitation of a solid, control of the above factors concerning the aqueous environment is very essential. They can further be used as valves to control and end any chemical reaction that leads to nucleation of a specific particle in a solution during a chemical precipitation process and in order to have a nucleation that leads to the formation of a specific particle starts when a supersaturation state is reached. Supersaturation is a state where a solution usually contains more dissolved material than could dissolve in a solvent under normal conditions. Supersaturation state is frequently used in the industry and it allows the manufacturing of many products, a common example is the solid candy that is manufactured from supersaturated sugar—water solutions. The current interest of researchers is not in random precipitation, but a precipitation that results in acceptable monodispersed NPs. Such a process was achieved and reported as early as in 1941 [9] for aerosols and was extended for sulfur sols few years later [10,11]. However, the theory that describes this homogenous precipitation of monodispersed particles was developed by LaMer et al. published in 1950 [12,13].

By presenting and studying results on the reaction of an acid and thiosulfate, LaMer presented the theory of homogenous formation of dispersed particles through chemical precipitation. LaMer and his co-workers have shown how monodispersed sols can be prepared through the dilution approach. They studied the mechanism of formation based on the relationship between the initial concentration of sulfur and the volume of the added water. Following the relative concentration variation LaMer et al. suggested a qualitative explanation for the mechanism of formation of monodispersed particles in sols. This qualitative explanation is schematically represented in Fig. 5.2 [12,13]. In the studied case the concentration of sulfur molecules increases slowly to form the dissolved sodium

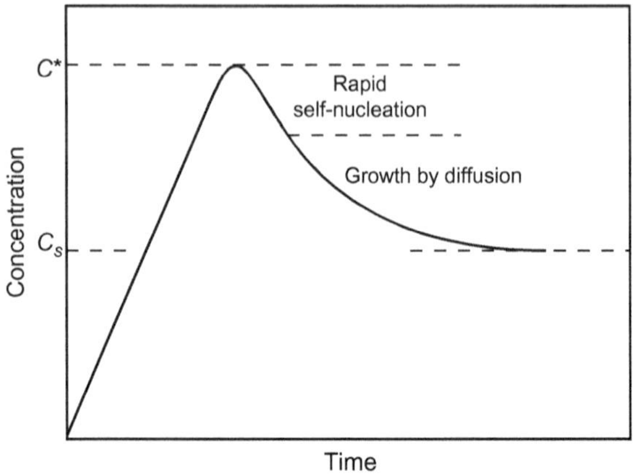

Figure 5.2 Schematic representation of the concentration with time of a particle formatting solute species generated in situ before and after self-nucleation. Here C_s represents solubility concentration, while C^* represents critical supersaturation of the particle forming solute species [12].

thiosulfate ($Na_2S_2O_3$). The increase of the concentration of sulfur will continue until a supersaturation state is reached. At supersaturation concentration, the nucleation of solid sulfur will prevail. At this stage, the nuclei of the particle start to grow through diffusion. Hence, the bottleneck for the growth of synthesizing uniform dispersed relatively small-size particles lies in the controlled generation of solutions that only precipitate leading to one burst of nuclei separately existing [8,12,13]. The species of the nuclei should then continue to be formed at a rate that allows their removal by diffusion onto the existing particle to avoid the formation of a secondary nucleation that might start to form [8]. This will lead to the situation that the original nuclei grows into a uniform dispersed particle of relatively small size. If the initial concentration of the nutrients that lead to the chemical precipitation reaction is not low enough, the rate of production of the precipitate will be relatively high, and the concentration of that leads to the nucleation, which leads to the precipitation of small-size monodispersed particles. Hence, large formed particles leading to polydispersed solution will be the result. Hence, the initial concentration is also vital for the formation of small-size monodispersed particles [12,13].

Nowadays and based on the same theory developed by LaMer et al. [12,13] and discussed above, precipitation is a widely used method for the synthesis of NPs with deep sub-100 nm size. These deep sub-100 nm particles are usually precipitated within a fluid solvent. An inorganic salt, such as a nitride, chloride, acetate, etc. is dissolved in water to a specific diluted concentration. In this case, the metal ions will exist as hydrates. Then this metallic hydrate is added to a basic solution, for example, NaOH. The mixture will be left for a certain period of time, usually a few hours. The hydrolyzed metallic ions will then condense as a precipitate. This will usually be accompanied by a color change of the mixed solution. Then this precipitate will be separated, washed, and dried. The resulting dried product will be in the form of a powder consisting of small agglomerated NPs. In general, the precipitation method is not a well-controlled method when considering the reaction kinetics, the solid phase nucleation, and the crystal growth. Therefore NPs synthesized by the precipitation method might have a wide range of size distribution, varied morphology, and are agglomerated [14,15]. To avoid these shortcomings, researchers have developed different approaches to achieve morphology and size control of well-separated NPs. The phenomena that lead to non-controlled size distribution and morphology variation are the thermal coagulation and Oswald ripening. Oswald ripening is a phenomenon that is observed when solid particles are available in solutions, that is, liquid sols [16]. Oswald ripening describes the change of an inhomogeneous structure of the solid particles over time. Both the thermal coagulation and Oswald ripening are avoided using a double layer repulsion of crystallites with nonaqueous solvents at a lower temperature of synthesis [14,15]. The nonaqueous solvent is usually added to the parent solution before precipitation takes place, and this surfactant is used so that the precipitated NPs maintain a distance between them [17]. Further rigorous stirring, which is often used nowadays, during precipitation, will be helpful to reduce or even diminish thermal coagulation leading to rather uniform size distribution as elaborated by the examples to follow.

Most of the recent research efforts using the chemical approach in synthesizing nanostructures have been concentrated on metal oxide nanostructures. The next important two families are compound semiconductor and metal NPs. The importance of metal oxide NPs is due to many reasons among which is the easy synthesis due to the availability of metal precursors and the variety of properties that can be utilized for many different applications. The fact that the metal oxide family is the most researched for many years and up to now is clear from the relatively large number of publications. Hence, most, but not all, of the examples that will be described in this chapter and in the next chapter will cover different metal oxide nanostructures.

Using 99.0% analytical grade chemicals, nickel oxide (NiO) NPs were prepared using nickel nitrate hexahydrate ($Ni(NO_3)_2 \cdot 6H_20$) and sodium hydroxide (NaOH) [18]. To study the effect of the surfactant on the size distribution of the precipitated NPs, three different surfactants were used. These were polyvinylpyrrolidone (PVP, M5000), polyethylene glycol (PEG, M5000), and cetyl trimethyl ammonium bromide (CTAB). In these experiments, two separate solutions of nickel nitrate and another of sodium hydroxide were prepared. For the first solution, 8.7 g of nickel nitrite was dissolved in 60 mL of deionized water, and 30 g of sodium hydroxide was dissolved in 150 mL of deionized water. Then 1 g of one surfactant was used in a separate experiment to investigate the effect of these surfactants on the variation of the size distribution of the synthesized NiO NPs. The surfactant was first added to the sodium hydroxide solution. Then the nickel oxide hexahydrate solution was added dropwise into the surfactant and the sodium hydroxide mixture [18]. The mixture was then stirred using a magnetic stirring apparatus. All processes were performed at room temperature. The resulting light green solution was then filtered and washed with deionized water many times (5–10), and then it was dried at 50°C. Fig. 5.3A–C shows the scanning electron microscope (SEM) images of the three types of NiO NPs synthesized using the three different surfactants. All three figures (A–C) indicated that NPs have been successful synthesized. In all three cases, the NP range was between 25 and 65 nm. For the PVP the maximum size was 55 nm. While for the PEG the NPs' variation was observed to be least between the smallest and the largest NP. While for the other two surfactants, that is, the PVP and CTAB, the variation was larger as shown in Fig. 5.4. Transmission electron microscope (TEM) image (not shown here) of the nickel oxide NPs synthesized using PEG as a surfactant has indicated uniform NPs with the narrowest size distribution. It is also noted that the agglomeration was quite weak and the NPs were separated from each other. As a conclusion, it was observed that PVP and PEG are the best two surfactants for achieving smaller size distribution with weak agglomeration for NiO NPs using the precursors described. Nevertheless, the surfactants used to avoid agglomeration are to be removed after synthesis if pure NPs are targeted. The preparation of monodispersed small-size passivated NPs is important as it allows the researchers to distinguish their exotic novel properties inherent to the nanoscale nature from other properties associated with heterogeneities or polydisperse related property (see Fig. 5.5) [19]. As it might be difficult to obtain identical NPs, nowadays,

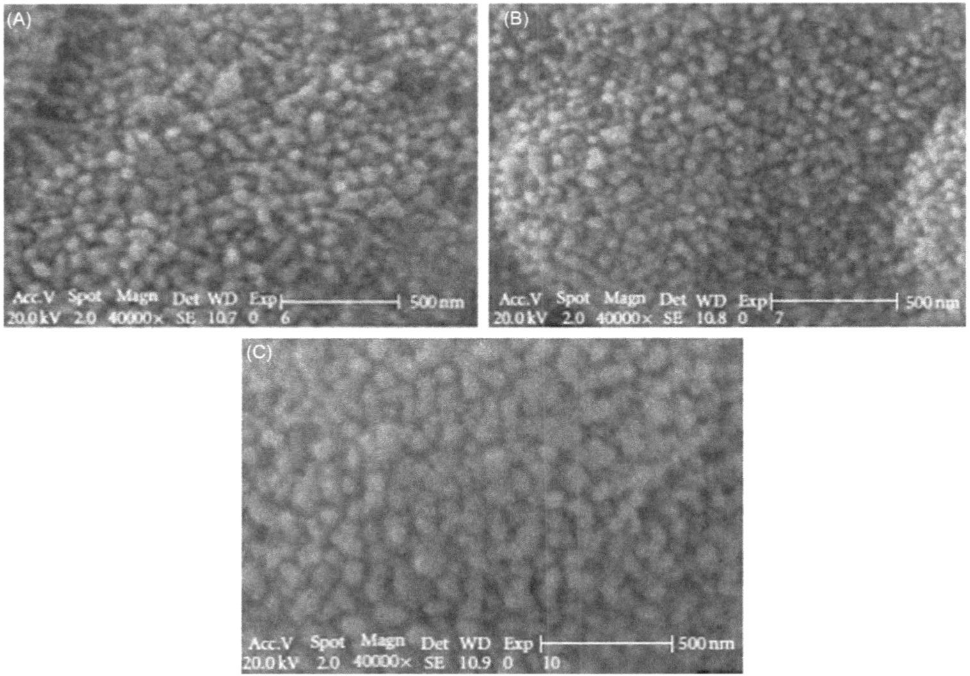

Figure 5.3 Scanning electron microscope images of NiO synthesized by the precipitation method using (A) PVP, (B) PEG, and (C) CTAB, as a surfactant [18]. *CTAB*, Cetyl trimethyl ammonium bromide; *PEG*, polyethylene glycol; *PVP*, polyvinylpyrrolidone.

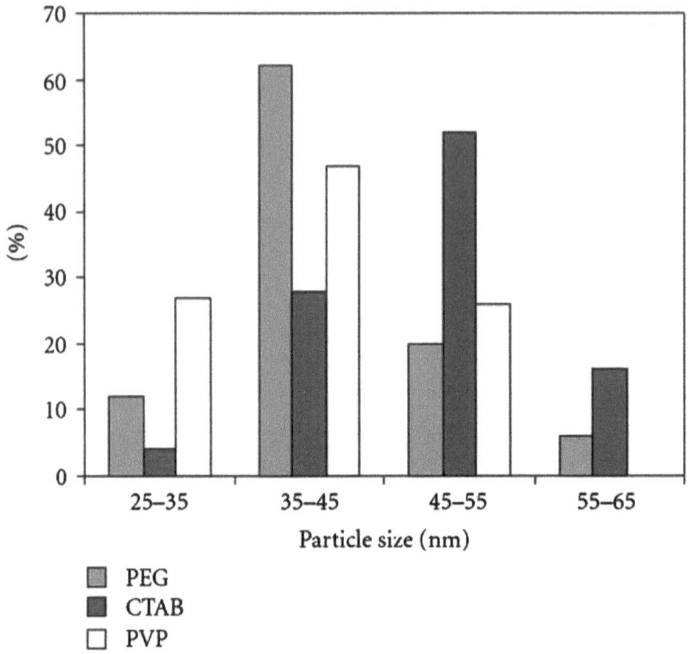

Figure 5.4 Effect on surfactant on the size and its distribution [18]. *CTAB*, Cetyl trimethyl ammonium bromide; *PEG*, polyethylene glycol; *PVP*, polyvinylpyrrolidone.

monodispersed NPs are a term referring to those NPs with a standard deviation $\sigma \leq 5\%$ in diameter variation [20]. In addition to this, NPs must, beside uniformity of size and shape, be of well-defined crystalline core and possess controlled surface chemistry [19]. Size uniformity of NPs diameter having standard deviation $\sigma \leq 5\%$ in diameter has been the topic of many research laboratories. The Ostwald ripening mentioned above has been utilized to synthesize size series NPs [21]. This is achieved by removing portions of the reaction precipitate after different durations form the start of the precipitation reaction. As can be seen from the cartoon presentation of LaMer classical growth of monodisperse NPs (discussed above and shown also in Fig. 5.2), the Ostwald ripening effect leads to the synthesis of different size NPs as reaction time evolves. Using this approach Murray et al. have managed to utilize a one-step precipitation reaction to synthesize CdE (E = S, Se, Te) high degree monodisperse NPs having sizes ranging between 15 and 115 Å [21]. The method is based on the pyrolysis of organometallic reagents using injection into a hot solvent [21]. Such a process will allow a controlled growth of

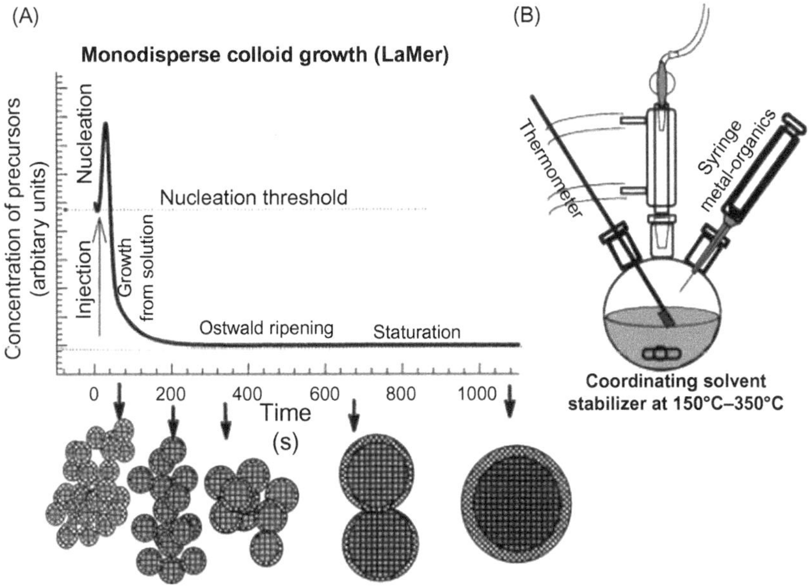

Figure 5.5 (A) Cartoon depicting the stages of nucleation and growth for the preparation of monodisperse NCs in the framework of the LaMer model. As NCs grow with time, a size series of NCs may be isolated by periodically removing aliquots from the reaction vessel. (B) Representation of the simple synthetic apparatus employed in the preparation of monodisperse NC samples [19]. *NCs*, Nano-crystals.

macroscopic quantities of size-selected NPs through precipitation. In fact, such a process will isolate portions of the growth solution having a narrow size distribution (<5% rms in diameter).

The experiments of the synthesis of the monodisperse CdE (E = S, Se, Te) NPs having a wide size series as described by the authors are as follows [21]:

5.2.1 Chemicals

All manipulations involving alkylcadmium, silylchalconides, phosphines, and phosphine chalconides were carried out using standard airless procedures. Tri-*n*-octylphosphine (TOP) and bis(trimethylsilyl) sulfide [$(TMS)_2S$] were used as purchased from Fluka. Electronic grade (99.99 + %) selenium and tellurium shot were purchased from Alfa. Anhydrous methanol, 1-butanol, pyridine, and hexane were purchased from a variety of sources. Tri-*n*-octylphosphine oxide (TOPO) was purchased from Alfa and purified by distillation, retaining the fraction transferred between 260°C and

300°C at ~1 Torr. Dimethylcadmium [Me_2Cd] was purchased from Organometallics Inc. and purified by filtration (0.25 μm) and vacuum transfer. Bis(trimethylsilyl)selenium [$(TMS)_2Se$] and bis(*tert*-butyldimethylsilyl)tellurium [$(BDMS)_2Te$] were prepared via as described in [22,23] and stored at −35°C in a drying box. Appropriate masses of selenium and tellurium shot were dissolved directly in sufficient TOP to produce 1.0 M stock solutions of trioctylphosphine selenide [TOPSe] and trioctylphosphine telluride [TOPTe] [24].

5.2.2 Method 1

The typical preparation of TOP/TOPO-capped CdSe nanocrystalline NPs is as follows [21]: 50 g of TOPO is dried and degassed in the reaction vessel by heating to ~200°C at ~1 Torr for ~20 minutes, with periodic flushing with argon. The temperature of the reaction flask is then stabilized at ~300°C under ~1 atm. of argon. Solution A is prepared by adding 1.00 mL (13.35 mmol) of Me_2Cd to 25.0 mL of TOP in the drying box. Solution B is prepared by adding 10.0 mL of the 1.0 M TOPSe stock solution (10.00 mmol) to 15.0 mL of TOP. Solutions A and B are combined and loaded into a 50-mL syringe in the drying box. The heat is removed from the reaction vessel. The syringe containing the reagent mixture is quickly removed from the drying box and its contents delivered to a vigorously stirring reaction flask in a single injection through a rubber septum. The rapid introduction of the reagent mixture produces a deep yellow/orange solution with an absorption feature at 440−460 nm. This is also accompanied by a sudden decrease in temperature to about 180°C. Heating is restored to the reaction flask and the temperature is gradually raised to 230°C−260°C. Aliquots of the reaction solution are removed at regular intervals (5−10 minutes) and absorption spectra taken to monitor the growth of the crystallites. The best quality samples are prepared over a period of a few hours of steady growth by modulating the growth temperature in response to changes in the size distribution as estimated from the absorption spectra. The temperature is lowered in response to a spreading of the size distribution and increased when growth appears to stop. When the desired absorption characteristics are observed, a portion of the growth solution is transferred by cannula and stored in a vial. In this way, a series of sizes ranging from ~15 to 115 Å in diameter can be isolated from a single preparation process. CdTe nanocrystalline NPs are prepared by Method 1 with TOPTe as the chalcogen source, an injection temperature of ~240°C, and growth temperatures between ~190°C and ~220°C [21].

5.2.3 Method 2

A second route to the production of CdE (E = S, Se, Te) nanocrystalline NPs replaces the phosphine chalconide precursors in Method 1 above with $(TMS)_2S$, $(TMS)_2Se$, and $(BDMS)_2Te$, respectively. Growth temperatures between ~290°C and ~320°C were found to provide the best CdS samples. The smallest (~12 Å) CdS, CdSe, and CdTe species are produced under milder conditions with injection and the growth is carried out at ~100°C [21].

5.2.4 Isolation and purification

The following method has been reported by the authors in [21]:

A 10-mL aliquot of the reaction solution is removed by cannula and cooled to ~60°C, slightly above the melting point of TOPO. Addition of 20 mL of anhydrous methanol to the aliquot results in the reversible flocculation of the nanocrystallites. The flocculate is separated from the supernatant by centrifugation. Dispersion of the flocculate in 25 mL of anhydrous 1-butanol followed by further centrifugation results in an optically clear solution of nanocrystallines and a gray precipitate containing by-products of the reaction. Powder X-ray diffraction (XRD) (PXRD) and energy dispersive X-ray analysis indicate these by-products consist mostly of elemental Cd and Se. This precipitate is discarded. Addition of 25 mL of anhydrous methanol to the supernatant produces flocculation of the crystallites and removes excess TOP and TOPO. A final rinse of the flocculate with 50 mL of methanol and subsequent vacuum drying produces ~300 mg of free-flowing TOP/TOPO-capped CdSe nanocrystalline. The resulting powder is readily dispersed in a variety of alkanes, aromatics, long-chain alcohols, chlorinated solvents, and organic bases (amines, pyridines, furans, phosphines) [21].

5.2.5 Size-selective precipitation

The size separation process was reported by the authors to be achieved as follows [21]:

Purified nanocrystals are dispersed in anhydrous 1-butanol forming an optically clear solution. Anhydrous methanol is then added dropwise to the dispersion until opalescence persists upon stirring or sonication. Separation of supernatant and flocculate by centrifugation produces a precipitate enriched with the largest crystallites in the sample. Dispersion of the precipitate in 1-butanol and size-selective precipitation with methanol is repeated until no further sharpening of the optical absorption spectrum is noted. Size-selective precipitation can be carried out in a variety of solvent/nonsolvent pairs, including pyridine/hexane and chloroform/methanol.

Fig. 5.6 demonstrates size-dependent precipitation outcome. The different spectra show the initial and the consequent different sizes after

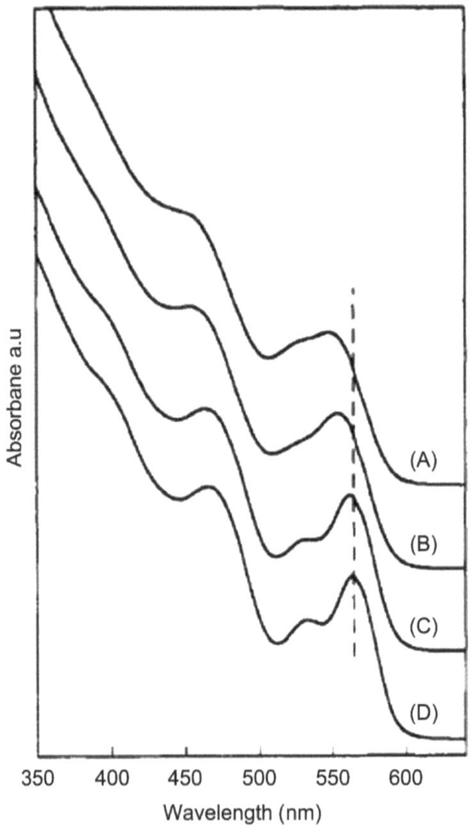

Figure 5.6 Example of the effect of size-selective precipitation on the absorption spectrum of ~37 Å diameter CdSe nanocrystallines. (A) Room temperature optical absorption spectrum of the nanocrystals in the growth solution before size-selective precipitation. (B) Spectrum after one size-selective precipitation from the growth solution with methanol. (C) Spectrum after dispersion in 1-butanol and size-selective precipitation with methanol. (D) Spectrum after a final size-selective precipitation from 1-butanol/methanol [21].

the selective precipitation was utilized for CdSe nanocrystals of deep sub-100 nm sizes. Fig. 5.6A shows the absorption spectrum before size-dependent precipitation is used. In the spectrum shown in Fig. 5.6B the nanocrystals absorption is shown after the addition of methanol. While the spectrum shown in Fig. 5.6D displays the absorption of the final size precipitation leading to CdSe nanocrystals of a 37 Å ± 10% sizes after the dispersion in 1-butanol. The initial CdSe nanocrystals were having a size of ~400 nm, and as can be seen in Fig. 5.6, the size was reduced down to 37 Å. Using this method it was possible to synthesize different II–IV compound semiconductor nanocrystals ranging in size from 12 Å and up to 115 Å in a single precipitation reaction [21]. However, this size-dependent technique suffers from the fact that it cannot be scaled up as the process is performed in a rather small chemical flask, and hence leads to synthesis of a limited amount of the required precipitate [25]. It is also to be noted that the above size-dependent approach cannot be scaled up as it is performed in a glass vessel, and if a larger vessel is used, the control temperature and homogenous mixing are not easily accomplished, especially for raid reactions like those utilized for precipitation. Despite the fact that the method presented above can provide high-quality CdSe and other II–IV compound semiconductor nanocrystals, a large batch preparation approach is of interest since the above size separation reported method can only yield 5–50 mL. This is due to the potential of these II–IV nanocrystals in many applications. It is well known that when these II–IV compound semiconductor nanocrystal have a size smaller than the Bohr radius, then their bandgap widens and consequently their florescences shift toward shorter wavelengths leading to potential applications, for example, fluorescence tags, light-emitting diodes, optical memories. Due to the high demand for these II–IV compound semiconductor nanocrystals with a deep sub-100 nm size, many efforts to develop scaled up processes have been published, not only for II–IV based nanocrystals but also for other NPs. Control of the temperature was the most important subject of study. Among these efforts is the development of a microreactor to control the temperature [26–28]. The use of a microreactor has successfully led to accurate temperature control for precipitation synthesis, however, still the scaling up of the process remained a challenge. This was partially solved by the use of a continuous flow reactor shown in Fig. 5.7 and developed by Nakamura et al. [29].

In general, an efficient liquid phase approach for NPs synthesis is determined basically on the range of the synthesized NPs size. The narrower the

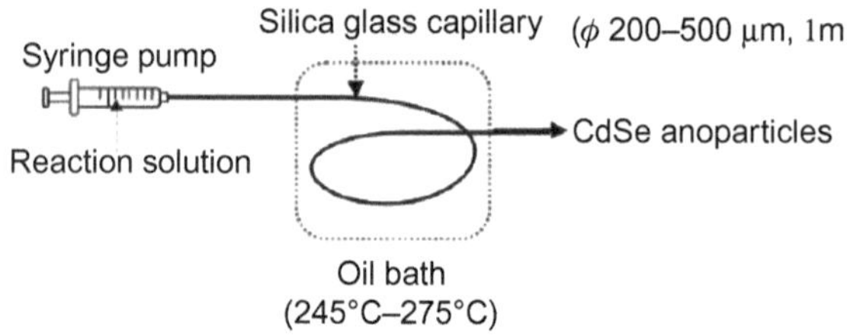

Figure 5.7 The continuous flow reactor developed by Nakamura et al. [29].

range, the more efficient is the process. This criterion is based on an obvious fact; which is that the size of a NP is critical for the expected properties. Hence, a relatively narrow range assembly of NPs will yield a well-defined property, while a relatively wide range size distribution of NPs can lead to undefined properties that are not producible. A mathematical two module model that describes the NPs' distribution dynamics and control can be found in [30].

It is important to mention that as a special case of precipitation; coprecipitation is also another process that occurs and is commonly used for the synthesis of a variety of NPs. Coprecipitation occurs when a foreign solute is captured within a precipitate with a stoichiometric configuration [31]. To illustrate the coprecipitation let us consider a water solution containing cation A^+ and anion B^-. When the anion and cation interact, they will lead to AB compound. Now if beside the cation A^+ and anion B^-, we have another two cations C^+ and D^+, when the two main cation and anion interact to form AB, a small fraction of the two foreign cations will be introduced and the result will be $(A_xC_yD_z)B$. In this case the values of y and z are quite small compared with x and the presence of the cations C and D is quite low and they are considered traces. At low temperature the miscibility of such foreign elements is decreased, and it is only limited to a few mole percentage fractions and hence can be ignored [31]. Nevertheless, the terminologies precipitation and coprecipitation are used today indistinguishably.

5.3 Low-temperature aqueous chemical synthesis of nanostructures

Nowadays the most popular methods for growing a wide variety of nanostructures are the low-temperature aqueous chemical methods. There are many reasons for making such methods attractive tor researchers, among them are the simplicity, wide range of nanostructures possible, low cost, no need for complex instrumentation, etc. This method is commonly referred to as low-temperature hydrothermal synthesis. By hydrothermal it is meant that the temperature and pressure are kept under control.

In many cases a closed vessel called the autoclave is used, as will be elaborated soon. The final product size and morphology when using the low-temperature hydrothermal synthesis is well investigated by considering the nucleation mechanism of the initial NPs [32]. Monitoring the initial NP nucleation is often studied following the description of conventional solid-state reaction equations. Here we consider the case of ferroelectric nanostructures kinetics of growth. Ferroelectric materials are crucial for many technological applications, and when utilizing their nanostructures' forms, scaling down ferroelectric devices will be of great benefit. We here follow and consider the growth description for $BaTiO_3$, which is a popular ferroelectric material. Many researchers describe the nucleation mechanism when utilizing the hydrothermal synthesis of the ferroelectric $BaTiO_3$ nanostructures by using the Johnson–Mehl–Avrami equation [33,34]. The Johnson–Mehl–Avrami equation describes the transformation of matter from one phase to another phase at constant temperature. Hence, it is useful in studying kinetics of crystallization of matter when considering the low-temperature aqueous chemical hydrothermal synthesis since it is a process that concerns the formation of a solid phase crystal from a liquid phase. The Johnson–Mehl–Avrami equation states [33,34]:

$$f = 1 - \exp(-k(t - t_0)^n) \tag{5.1}$$

where f is the fractional extent of the reaction as a function of time after the first appearance of the initial phase (t_0), k is the rate constant, while n is an exponent related to the mechanism of growth [33]. As mentioned above, and although this model is based on solid-state reaction it has been successfully used to accurately describe low-temperature aqueous chemical hydrothermal kinetics and the mechanism of the synthesis of a variety of nanostructures [35–37]. This is usually achieved using in situ monitoring of the growth kinetics. The group of nanostructures of the family of transition metal antimony oxides, that is, MSb_2O_4 (M: Mn, Fe, Co, Ni, Zn), has different applications, among which is the use as an anode for Lithium batteries. Using the hydrothermal low-temperature synthesis, different growth runs for a temperature range between 135°C and 300°C were performed to synthesize $CoSb_2O_4$ [36], and the model was successfully used to follow the growth kinetics. By applying the Johnson–Mehl–Avrami model, both the reaction mechanism as well as the activation energy can be extracted. The reaction extent $f(t)$ in Eq. (5.1) is usually plotted as a function of the time as the appearance of the first nanocrystalline $(t - t_0)$ [36]. Subsequently the data is modeled according to Eq. (5.1). Here the extent of the reaction, that is, $f(t)$ is modeled as $V(t)/V_{inf}$, where V_{inf} is the final stable nanocrystal volume at a specific temperature [36]. Here for the case of the $CoSb_2O_4$, the assumption was that the initial morphology of the crystal is spherical [36]. The two parameters in the Johnson–Mehl–Avrami equation, namely k and n, that is, the rate constant and the mechanism parameter respectively, can be determined. For n value between 0.54 and 0.62 the reaction follows a diffusion-controlled path, while when n is between 1.0 and 1.24 the reaction is zero order, first-order, or a phase-boundary controlled mechanism, and when n is between 2.0 and 3.0, then the growth corresponds to a nucleation and growth-controlled mechanism [38].

For the low-temperature aqueous chemical hydrothermal synthesis of $CoSb_2O_4$, the analytical grade precursors used were purchased from Sigma Aldrich [36]:

Two different synthesis experiments were performed. These were either for in situ characterization or ex situ characterization. For the in situ characterization the following precursors were used:

NaOH, HCl, and Co-$(CH_3COO)_2 \cdot 4H_2O$ ($\geq$98.0%,) used as the Co sources. For the Sb source, a $SbCl_3$ in 4 M HCl and Co $(CH_3COO)_2 \cdot 4H_2O$ in a deionized solution was utilized. To adjust the pH NaOH was used. By adding NaOH the solution color was changed from red to dark blue. For the synthesis a 2.5 mL of 1.00 M Co-$(CH_3COO)_2 \cdot 4H_2O$ solution was mixed with 2.5 mL of 1.33 M $SbCl_3$ solution. Then this was followed by the addition of 3.0 mL of 8 M NaOH. For the optimization of the growth, the ratio of the Sb to Co was either 1.33 or 2.0. The variation of the ratio of the Sb to the Co was performed to obtain insight into the influence of the variation of this ratio on the nanocrystals' formation. For more information on the synthesis of the present $CoSb_2O_4$, the reader is referred to [36]. The in situ characterization was performed inside an in situ PXRD equipment. The details of this setup are described in detail and can be found in [39].

For the ex situ characterization, the precursors below were used.

Here larger quantities of the synthesized material were needed. 10 mL of 0.667 M $Co(CH_3COO)_2 \cdot 4H_2O$ was mixed with 10 mL 1.334 M $SbCl_3$ that was dissolved in a 4 M HCl. While using a magnetic stirring, 16 mL of 8.0 M NaOH was added to the first mixture. This mixture was kept inside an autoclave and the temperature was kept at the required value for 53 hours [36]. Usually for ex situ investigations the reaction is usually interrupted at certain durations and the products are then investigated to follow the reaction kinetics. Different ex situ studies can be found in [40–43].

Fig. 5.8A–D shows the dependence of the extent of reaction ($f(t) = V(t)/V_{\text{inf}}$), and the dependence on ($t - t_0$) for four different temperatures ranging between 135°C and up to 220°C. By utilizing the Johnson–Mehl–Avrami equation and using the fitting technique, both the rate constant (k) and mechanism-related parameter (n) can both be obtained. The values of n were found to vary from 0.9 up to 0.4, as can

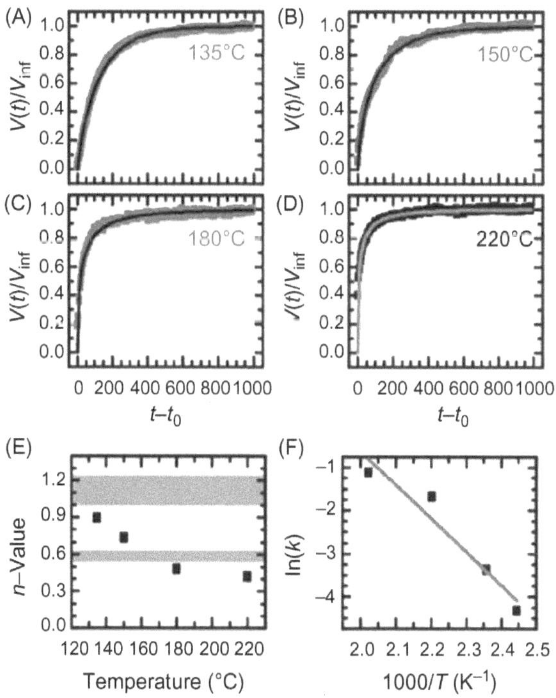

Figure 5.8 (A–D) Johnson–Mehl–Avrami plots for syntheses at (A) 135°C, (B) 150°C, (C) 180°C, and (D) 220°C in the nonstoichiometric case. (E) n Values obtained from the fits at different synthesis temperatures with gray areas indicating physically meaningful n values. (F) Arrhenius lot.

be seen in Fig. 5.8E. As the temperature was increased the value of n was decreased. This implies that the reaction mechanism was found to change from being either zero- or first-order kinetics to a diffusion-controlled reaction mechanism [36]. Nevertheless, with few experimental data points, a concrete conclusion is not convincing and a mixture of more than one reaction mechanism is not excluded. The reaction activation energy (E_a) can be found by plotting the Arrhenius equation which reads:

$$k = A\exp\left(\frac{E_a}{RT}\right) \tag{5.2}$$

where T is the absolute temperature, R is the universal gas constant, k is the rate constant, and E_a is the activation energy. By plotting ln(k) versus 1000/T, the slope will give the activation energy, as shown in Fig. 5.8F. In this case for $CoSb_2O_4$ the activation energy was found to be 65(12) kJ/mol [36]. This value is close to other values obtained from different metal oxides synthesized by the low-temperature hydrothermal aqueous chemical synthesis. There are much more published data on the successful use of the Johnson–Mehl–Avrami equation for modeling and fitting the experimental results on different nanostructures synthesis with the aim of revealing the reaction kinetics. Further theoretical computer modeling was applied to investigate the accuracy of the Johnson–Mehl–Avrami equation [44]. It was found that the Johnson–Mehl–Avrami equation provides an accurate description of the phase transformation and the growth kinetics.

Using the hydrothermal low-temperature aqueous chemical synthesis different morphologies of nanostructures can be demonstrated. By tuning the growth parameters, a specific nanostructure morphology with different geometrical dimension can be obtained. Zinc oxide (ZnO) nanostructures are the most researched material. The reason is due to the excellent manifold properties of this material, in addition to the fact that the family of ZnO nanostructures is the richest family when considering the different possible morphologies. Zinc oxide is a II–VI wide bandgap (E_g) semiconductor, having $E_g = 3.4$ eV at room temperature, and at the same time it has a relatively large excitons binding energy of 60 meV [44]. Zinc oxide has a hexagonal wurtzite crystal structure with no color when undoped. In addition to its excellent optoelectronic properties—ZnO is piezoelectric—it is also biocompatible and biosafe [44]. Zinc oxide also possesses a large number of point defects that occupy energy levels within the bandgap.

The appearance of these point defects is due to both intrinsic or extrinsic sources. These mid-bandgap states emit all the visible spectrum colors and hence ZnO is capable of emitting intrinsic white light [45]. This yields a photoluminescence (PL) spectrum that is characterized by two main peaks. The first is a sharp peak centered at around 380 nm and is called the UV band edge emission, while the second is a broad peak extending between 400 and 750 nm and is centered at around 550 nm. This is why this peak is commonly referred to as the "green emission peak," since it is centered at the green emission wavelength. Despite the rewarding properties of ZnO and the fact that it has been known to scientist since the 1940s [46], the research on this excellent material has only been active during the last decade. The reason that hindered the realization of the excellent properties of ZnO was due to the fact that to date there is no approach to obtain p-type doping of this material. Nevertheless, with the appearance of the rich family of a variety of morphologies of ZnO nanostructures, and the fact that these nanostructures have a small footprint that facilitates their synthesis on any substrate with no need for lattice mismatch, the research on utilizing ZnO for different application has intensified dramatically. Fig. 5.9 displays a typical PL spectrum of ZnO nanomaterial [47]. Adding to this the different low-temperature methods that are available for obtaining these different ZnO morphologies has led to them being synthesized on any substrate, even those with amorphous natures [48,49]. Fig. 5.10 shows typical ZnO nanowires (NWs) grown by the low-temperature aqueous chemical approach

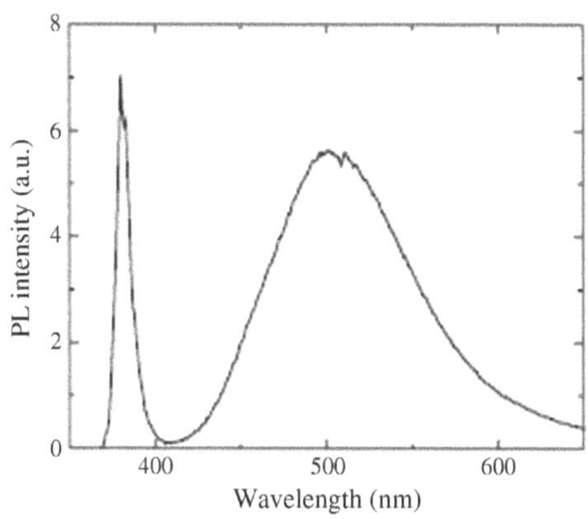

Figure 5.9 Typical photoluminescence room temperature spectrum of ZnO material [47].

Figure 5.10 (A)-(D) Different magnification scanning electron microscope images of ZnO nanowires grown on thin glass pipette by the low-temperature aqueous chemical synthesis method [50].

on relatively thin amorphous glass pipette surface [50]. Hence below we discuss some of the findings using the low-temperature chemical synthesis approach for ZnO nanostructures. As mentioned above, ZnO possesses the richest family of different morphology of nanostructures. Among the different morphologies, ZnO NWs and nanorods (NRs) are among the most interesting morphologies due to their one-dimensional nature that is beneficial for many optical properties.

In general, metal oxide nanostructures being of single metal oxide or composite metal nature are of extreme technological importance. This is due to the many possible future devices that can benefit from this class of material. As it is well known and for industrial realization the large-scale production of such nanomaterials/nanodevices should satisfy the criteria of low cost. Indeed the low-temperature aqueous chemical approach is a suitable candidate to adopt of realizing meal oxide nanostructures at low large-scale production cost. The concept of "purpose-built meal oxide nanomaterials" for the development of a new generation of smart materials and devices was introduced during 2000 [51]. To achieve this goal the control of the size and morphology of metal oxide nanostructures is of importance both from fundamental and industrial points of view. Hence, the successful well-controlled synthesis of three-dimensional meal oxide nanostructures is a critical step for the emergence of mature devices that suits the targeted technological application. Due to this, researchers have developed a plethora of new techniques [51]. It is to be noted that the aqueous chemical design of inorganic materials became attractive 30 years ago [52,53]. Generally, there are three different routes when considering the aqueous chemical design of solid inorganic materials, these are (1) strategies involving the traditional ceramic procedure and its variants as well as improvements directed toward decreasing diffusion distances, (2) novel reactions and methods to prepare known and sometimes new solids, and (3) soft chemical routes generally yielding new metastable solids [52]. For achieving device-quality reproducible three-dimensional metal oxide nanostructures, a control for the thermodynamics and kinetics of nucleation is necessary [52]. As nanostructures are characterized by relatively large surface area to volume ratios, most of the atoms are on the surface and hence controlling the atomic arrangements on the surface is crucial in determining the physical properties delivered by the nanomaterials synthesized. Hence, to control the physical properties, MM and design is the proper route to adopt [51]. The basic idea of "purpose-built nanomaterials," mentioned above is based on modeling and designing nanomaterials with specific size, morphology, and spatial distribution so specific properties are achieved [51]. For the proper design of a well controlled metal oxide nanomaterial, the free energy at the interface is the most crucial parameter [51,54].

A pioneering work on the low-temperature aqueous chemical synthesis of hematite (Fe_3O_4) and ZnO NWs material was published in 2001 [55–57]. When considering the homogenous nucleation of the solid phase of metal oxides it is to be noted that such process requires a relatively high activation energy barrier and that requirement leads to the promotion of heteronucleation from an energy point of view [56]. In this case the interfacial energy between the crystal and the solution is larger than the interfacial energy between the crystal and the substrate [56]. This will favor the heteronucleation of the crystal on the substrate at lower concentrations compared to the nucleation inside the solution. This heteronucleation on the substrate will proceed along the direction of crystallization of the material under investigation. Since divalent metal ions have very low tendency to precipitate in aqueous solutions through hydrolysis–condensation processes in neutral and acidic mediums, the synthesis of ZnO is designed to occur in an aqueous through thermal decomposition of Zn^{2+}–amino complex with reagent-grade chemicals [56]. In a typical synthesis experiment, a polycrystalline glass substrate is placed in a glass flask containing an equimolar (0.1 M) solution of zinc nitrate (Zn $(NO_3)_2$,$4H_2O$), and hexamethylenetetramine (HMTA), which has $C_6H_{12}N_4$ as its chemical formula. HMTA, also called methenamine, has been suggested to be used for the synthesis of the molecular assembly of crystals [58,59]. To build a complex nanostructure network the precise control of the placement of atoms is necessary. For the purpose of precise placement of atoms to form complex nanostructures it is possible that matured techniques like atomic force or tunneling electron microscopies can be utilized. However, such techniques are very slow and hence cannot be practical for industrial development. Hence, for industrial realization of nanostructures other building blocks must be developed; these building blocks can be considered as nanobricks that facilitate the demonstration of nanostructures networks. If self-assembly or position assembly techniques are to be combined with suitable building blocks, complex networks of a variety of nanostructures can be demonstrated. HMTA is emerging as a popular building block material for complex nanostructures networks development. HMTA, in the chemical synthesis of nanostructures, acts as a pH buffer in addition to its steric effect that promotes space for NRs/NWs growth at low temperature. If the temperature is relatively high, excess ammonia canceling the nonpolar chelation will be dominant and the synthesis will lead to rise-shape morphology [60]. This is experimentally seen by the common use of HMTA in low-temperature chemical synthesis of NRs/NWs of ZnO material. The abovementioned chemical flask containing a mixture of HMTA and zinc nitrate was inserted into a normal laboratory oven set at 9°C. The flask is then left for a period of time. How long the flask is left inside the oven depends on the size required from the specific experiment. After removal of the flask from the oven, the substrate is removed from the precursor solution and is washed by deionized water several times to remove any residuals residing on the surface, and the substrate is left to dry at normal room temperature [56]. Fig. 5.11 shows a typical SEM image of the produced ZnO NWs at two different magnifications (left). While on the right side of Fig. 5.11, the XRD pattern is displayed. As can be seen and as expected, this low-temperature aqueous chemical synthesis has yielded *c*-axis oriented

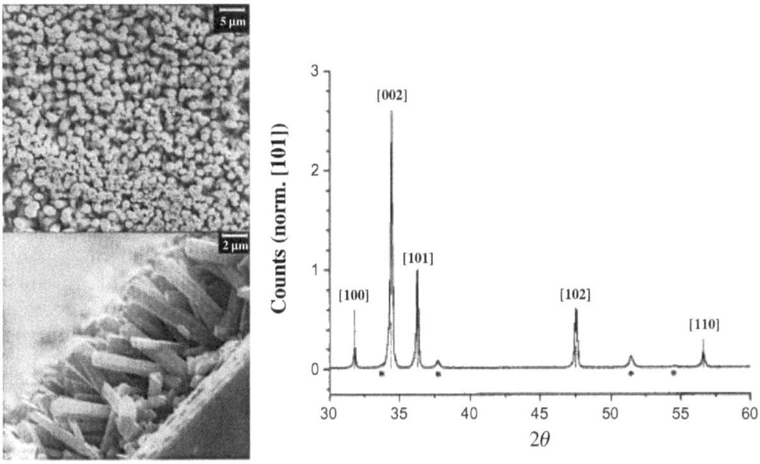

Figure 5.11 (Left) Typical scanning electron microscope images of ZnO nanowires at low (top) and high magnification (bottom). (Right) X-ray diffraction pattern of ZnO nanowires indicating *c*-axis oriented growth due to the high intensity of the 002 reflection [56].

NWs, as the *c*-axis is the favored growth direction of the ZnO crystal. The X-ray pattern shown in Fig. 5.11 indicates that the material is crystalline in nature. This implies that no postsynthesis annealing step is required. This is a positive fact as it means that soft substrates or other heat-sensitive substrates can be utilized for such synthesis process. This has a rather positive impact, as it can be a route for developing prototype devices on flexible and heat-sensitive substrates. Although the synthesis of ZnO NWs has been demonstrated using different synthesis precursors, the use of HMTA and zinc nitrate is very common. Using the combination of HMTA and zinc nitrate, the reaction leading to the synthesis of ZnO NWs is suggested to proceed as follows [61]:

$$(CH_2)_6N_4 + 6H_2O \rightarrow 6HCHO + 4NH_3 \tag{5.i}$$

$$NH_3 + H_2O \rightarrow NH_4^+ + HO^- \tag{5.ii}$$

$$Zn(NO_3)_2 \rightarrow Zn^{2+} + 2NO_3^- \tag{5.iii}$$

$$Zn^{2+} + 2HO^- \rightarrow ZnO(s) + H_2O \tag{5.iv}$$

The ZnO formation process, when using HMTA and zinc nitrate or any similar zinc salt, starts by the hydrolysis of the HMTA due to the heat and the result is NH_3 and formaldehyde (Eq. 5.1). Then the NH_3 reacts with H_2O to form NH_4^+ and HO^- (Eq. 5.2). At the same time the zinc nitrate produces Zn^{2+} when it is dissolved in H_2O (Eq. 5.3). Finally, solid ZnO is formed when Zn^{2+} reacts with $2HO^-$ (Eq. 5.4). This aqueous chemical synthesis of ZnO material was first reported in 1990 [62]. In this early work, the demonstrated structures were in the micrometer-size range, that is, microcrystals, and the first report on nanosize crystals was reported, as mentioned earlier 10 years after the first report [51,55,56]. In these early experiments, the formation of ZnO microcrystals was achieved by using the hydrolysis of zinc nitrate or zinc chloride in the presence of HMTA [62]. Depending on the synthesis parameters, for example, precursors concentration, pH, temperature, etc., different morphologies were obtained [62]. Fig. 5.12 shows some of the demonstrated ZnO microcrystals [62]. Hence, the ratio of the molar concentration of the HMTA to

Figure 5.12 (A–D) different ZnO microcrystals obtained by varying the synthesis parameters using zinc nitrate or zinc chloride together with HMTA demonstrated in the first report on aqueous chemical growth of ZnO [62]. *HMTA*, Hexamethylenetetramine.

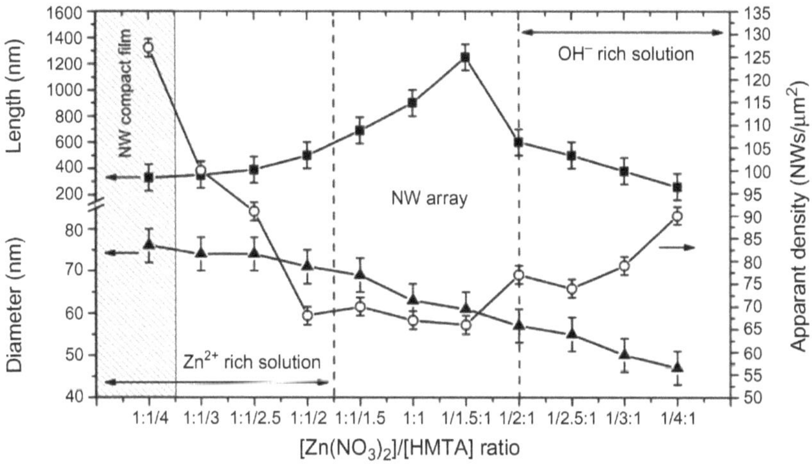

Figure 5.13 Evolution of the diameter (*triangle dots*), length (*square dots*), and apparent density (*circle dots*) of ZnO NWs as a function of the $[Zn(NO_3)_2]/[HMTA]$ ratio. *HMTA*, Hexamethylenetetramine; *NW*, nanowire.

the zinc precursor has a vital role of the morphology of the synthesized NWs/NRs. As shown in Fig. 5.13, the length of the NWs varies when varying the ratio of the $[Zn(NO_3)_2]/[HMTA]$. As shown the length of the ZnO NWs have increased from 300 to 1250 nm as the ratio is decreased from 4 to 0.66. As the ratio $[Zn(NO_3)_2]/[HMTA]$ is decreased further from 0.66, the length starts to decrease [60]. Hence, the longest NWs were 1250 nm and obtained at a nonequimolar ratio of 0.66, that is, 30 mM HMTA and 20 mM $Zn(NO_3)_2$. The ratio of the $Zn(NO_3)_2$ to the HMTA also has an influence on the width of the NWs. As can be seen in Fig. 5.13 the diameter of the ZnO NWs continue to decrease while the length increases when the ratio is varied from 4 to 0.25 [60]. This implies that the aspect ratio is increasing due to the increase of the length and decrease of the diameter as the ratio is decreased from 4 to 0.25 [60]. This implies that the sidewall growth, that is, radial growth is inhibited as the ratio of the $Zn(NO_3)_2$ to the HMTA is decreased from 4 to 0.25. The HMTA has been observed also to affect the apparent density of the ZnO NWs [60]. It was observed that as the ratio $[Zn(NO_3)_2]/[HMTA]$ is changed from 4 to 0.25, that is, increasing the HMTA, the density of the ZnO NWs is decreased by a factor of 1.4 [60].

After the pioneering work on ZnO NWs [51,55,56], and due to the multifold excellent promising properties of this material many reports were published. During the early work, the synthesis was a one-step process. Nevertheless, the vertical alignment of the produced one-step process was not satisfactory. A two-step process has then been developed just after the pioneering work mentioned above [63,64]. The developed two-step synthesis process starts with the spin coating of ZnO NPs of small size. Then it proceed with the earlier developed synthesis using the usual precursors, that is, $Zn(NO_3)_2$ and HMTA. The ZnO NPs is usually spun-coated on the surface of the substrate and the coated is repeated many times to insure homogenous spatial distribution of the ZnO NPs over the whole substrate surface [63,64]. The size of the ZnO NPs is usually between 5 and 10 nm in diameter. The substrate is subjected to annealing at 150°C after each spun-coating step. This annealing step will insure a good adhesion of the ZnO NPs to the substrate. The ZnO NPs were prepared using a solution of 0.03 M NaOH in methanol, which is then added dropwise to 0.01 M zinc acetate dihydrate in methanol, while the temperature is kept at 60°C with continuous stirring for 2 hours [65]. Such ZnO NPs solution can be stable for about 2 weeks and during this period, it can be used without fear of loss of stability [63]. Using this two-step ZnO NWs synthesis approach, well-aligned ZnO NWs on relatively large area substrate were demonstrated [63,64]. As shown in Fig. 5.14, a

Figure 5.14 ZnO nanowire array on a 4-in. (c. 10 cm) silicon wafer. At the center is a photograph of a coated wafer, surrounded by SEM images of the array at different locations and magnifications. These images are representative of the entire surface. Scale bars, clockwise from upper left: 2, 1 mm, 500, and 200 nm [63]. *SEM*, Scanning electron microscope.

rather vertically *c*-axis oriented ZnO NWs were demonstrated over a 4-in. wafer [6,63].

Beside the as $Zn(NO_3)_2$ and HMTA as the common nutrients for the low-temperature chemical synthesis of ZnO NWs, many researchers have used other precursors and achieved high-quality ZnO NWs. Among the utilized materials, zinc acetate-HTMA [66], zinc nitrate-thiourea-ammonium chloride-ammonia [67], zinc nitrate-triethanolamine-HCl [68], zinc acetate-sodium hydroxide-citric acid [69], to mention few. Beside the pH and the synthesis nutrients type or ratio, surfactants have been also utilized to control and modify the morphology of the grown nanostructure when using the low-temperature chemical methods [70−72]. In a typical experiment, ethylenediamine (5 or 10 vol.%) pH was adjusted to either 13.5 or 12. = for the two concentrations and is then added to zinc acetate dehydrate under vigorous treatment [71]. When the zinc acetate dehydrate was completely dissolved, a specific amount of pure analytical grade sodium hydroxide pellets were added so that an initial pH as shown in Table 5.1 is reached [71]. While the mixture was under continuous stirring, it was inserted in an oven that was kept at a temperature between 80°C and 100°C for a duration of 2 hours [71]. While stirring the mixture is left to cool down. The free standing precipitated white powder consisting of different ZnO nanostructures depending on the initial H value was collected and washed man times by deionized water and were left to dry under room temperature. Table 5.1 list all the archived ZnO nanostructures together with the starting synthesis parameters. As can be seen different homogenous and inhomogeneous ZnO nanostructures were synthesized by using the Ethylenediamine as a soft surfactant [71]. In addition to the modification of the morphology, the intrinsic properties of the grown nanostructures can also be tuned by adjusting the synthesis nutrients and other parameters [73−75]. This issue which is directly related to device applications will be discussed in details in Chapter 6: Emerging applications.

Due to the small footprint of the NWs and the nanostructures in general, the requirement of lattice mismatch substrate, which is necessary for thin films, is not required for the synthesis of nanostructures. This implies that nanostructures of excellent crystalline nature can be grown on any substrate, even those of amorphous nature [47,76]. Although most of the reports on the low-temperature chemical synthesis of ZnO NWs/nanostructures are performed at a temperature of around 90°C, it was shown that this synthesis could be achieved at even low temperatures and on

Table 5.1 Synthesis parameters and morphology of different ZnO nanostructures[a] [71].

Sample	E_n (%)	pH_i	pH_f	Morphology
ZnO-1	10	8.6	11.0	Elongated nanoparticles
ZnO-2	10	9.0	11.0	Flower like nanostructures
ZnO-3	10	10.0	11.0	Inhomogeneous nanorods
ZnO-4	10	10.0	12.0	Inhomogeneous nanorods
ZnO-5	10	11.0	12.0	Homogeneous nanorods of high aspect ratio
ZnO-6	5	8.6	11.0	Inhomogeneous nanorods
ZnO-7	5	9.0	12.0	Homogeneous nanorod bundles
ZnO-8	5	10.0	12.0	Homogeneous nanorods with low aspect ratio
ZnO-10	5	8.6	12.0	Defective nanoparticles, short nanorods, and rectangular nanosheets

[a]All of the sample were synthesized between 80°C and 100°C, with 2-h reaction time.

different substrates [76]. By lowering the synthesis temperature for ZnO NWs, a striking result was observed (see below). As ZnO NWs are of special interest due to their alignment and the potential of this properties for different optoelectronic applications, like wav guides, etc. we will below discuss the thermodynamics constrains that are to be fulfilled for the successful growth of ZnO nanostructures. As mentioned before, the deposition of a NP seed layer is a vital step for providing active nucleation for the nanostructure crystal. This seed layer is proven to play an important role on orientation and crystal quality of the specific nanostructure under synthesis.

The growth of the ZnO NWs is driven by the minimization of the Gibbs free energy. Due to nucleation, the Gibbs free energy change (ΔG) can be written as [77]

$$\Delta G = \frac{4}{3}\pi r^3 \Delta G_V + 4\pi r^2 \gamma \tag{5.3}$$

where r is the radius of the cluster, γ is the interfacial energy, and ΔG_V is the change of the Gibbs free energy per volume of the solid phase. Eq. (5.3) is composed of two terms that have an opposite influence and they affect the change of the Gibbs free energy due to nucleation.

The first, gives the energy gain required to create a new volume, while the second term gives the energy loss induced by the surface tension of the new interface. The change of the Gibbs free energy per unit volume of the solid phase ΔG_V is given by [77]:

$$\Delta G_V = \frac{k_B T}{V} \ln\left(\frac{C}{C_o}\right) \tag{5.4}$$

where k_B is Boltzmann's constant, C is the zinc concentration and C_o is the zinc solubility in the solution, and T is the temperature in Kelvin. By setting $d\Delta G/dr = 0$, the critical Gibbs free energy for nucleation ΔG^* can be obtained as [77]

$$\Delta G^* = \frac{16\pi\gamma^3}{3(\Delta G_V)^2} \tag{5.5}$$

The nucleation occurs when ΔG^* is minimum and this nucleation will be followed by uniform growth. Hence, the use of the ZnO NPs as a seed layer will lead to lower the thermodynamic barrier by acting as a nucleation site. Hence, the seed layer facilitates a uniform growth due to the lack of suitable nucleation sites. Without the use of a seed layer, no uniform growth can be achieved, and in addition, the use of the seed layer will increase the aspect ratio of the NWs [76]. The synthesis of well-aligned ZnO NWs was preceded by annealing the ZnO NPs coated substrate by a 250°C–300°C annealing step. This presynthesis annealing step was believed to essential as it decomposes the zinc acetate [78,79]. Nevertheless, other research findings provided results that show the fact that it is not necessary to perform presynthesis annealing step for the ZnO NPs seed coated substrates [76]. Indeed this findings is important due to the fact that it enables the use of soft substrates that cannot withstand such presynthesis annealing temperature range, for example, plastic, paper, etc. [76].

In Fig. 5.15A ZnO NWs grown on Ag coated plastic substrate is shown, the size was found to be between 120 and 235 nm. When the NWs were grown on Cu coated plastic substrate (Fig. 5.15B), was used the size of the ZnO NWs was ranging between 80 and 125 nm. While when using copper oxides (CuO and Cu_2O) substrate, the size was almost similar and was lying between 100 and 145 nm (shown in Fig. 5.15C and D). Finally, when using PEDOT/PSS as a substrate, the ZnO NWs were having an average diameter of about 100 nm [76]. For all the ZnO NWs grown the length was ranging between 1.2 and 2.0 μm. It have been

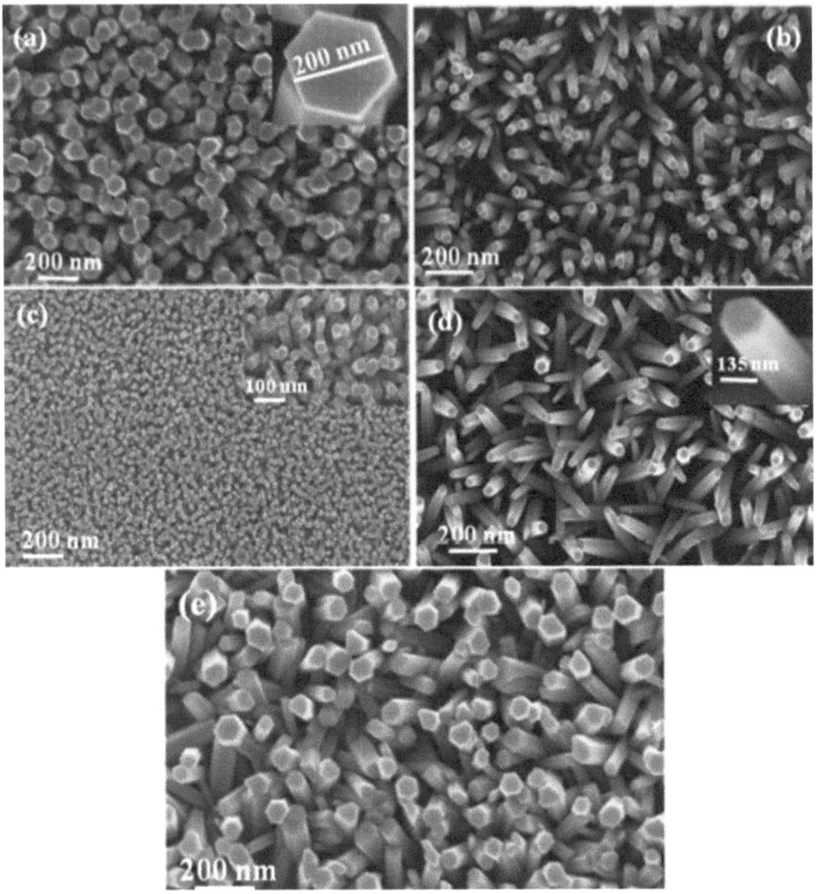

Figure 5.15 Scanning electron microscope top view of well-aligned ZnO NWs grown at 50°C on: (A) Ag coated plastic foil, (B) Cu coated flexible plastic, (C) CuO coated glass, (D) on Cu_2O glass substrate, and (E) on PEDOT/PSS [76]. *NW*, Nanowire.

reported that the synthesis temperature has a role in increasing the length of the ZnO NWs for the aqueous chemical synthesis, specifically, by increasing the temperature the length of the ZnO NWs is expected to decrease [79]. The justification of this was that by the effect of increasing the synthesis temperature on the critical radius of nucleation. At higher temperature, a competition between nucleation and growth takes will exists [76]. What will happen is that more zinc ions will be consumed in forming stable nucleation sites than initiating the growth. This will imply that thicker ZnO NWs will be achieved at higher temperature. The effect of the synthesis temperature on the aspect ratio, that is, the diameter and

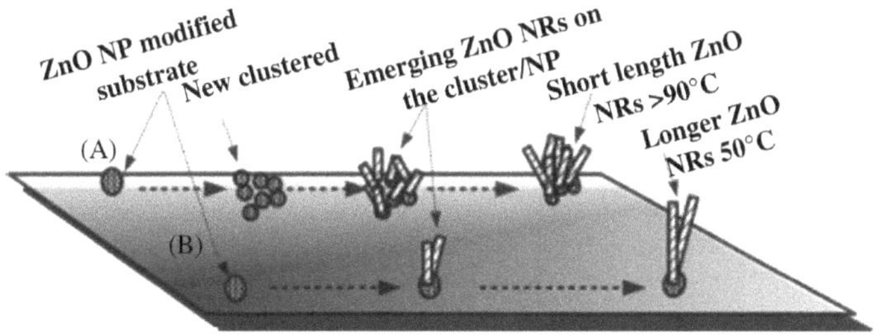

Figure 5.16 Schematic illustration of the nucleation/growth steps of ZnO NWs grown at (A) >90°C and (B) 50°C [76]. *NW*, Nanowire.

the length, can be discussed from a different point of view. In Fig. 5.16, two growth scenarios are shown; the first for 90°C and the second is for 50°C [76]. It was observed that as the synthesis temperature is lowered from 90°C to 50°C, the length of the ZnO NWs is increased, while the diameter is decreased, that is, the aspect ratio increases as the synthesis temperature is reduced. For the case of the synthesis at a temperature of 90°C, both the particle size as well as the radius of nucleation are reduced. Compared to the synthesis at 50°C, this will lead to increase the rate of nucleation as a consequence of the increase of the zinc solubility as the temperature increases [80,81]. This implies the fact that at relatively higher synthesis temperatures, completion between the axial vertical growth and nucleation on the substrate will lead to shorten the length of the ZnO NWs regardless of the increased density of the NWs [76]. For the synthesis at a temperature of 50°C or lower, the critical radius of nucleation remains almost constant and this will result in a longer ZnO NWs as observed [76]. Hence, by optimization for all synthesis parameters, that is, temperature, pH, nutrients concentration, etc., a control over the synthesized ZnO NWs can easily be achieved.

So far, we have discussed the synthesis of intrinsic nanostructures using the low-temperature chemical approach. However, doping such nanostructures is sometimes essential to further tailor their properties, being structural, optical, magnetic or of any other property of interest. Further, by using the low-temperature chemical synthesis it was also possible to successfully synthesize doped nanostructures, the results are published in many articles, for example, [82–88]. Among the interesting dopants for ZnO NWs is the family of the magnetic elements. Of special interest is to transfer ZnO NWs to be a diluted magnetic semiconductor as it can provide the route of room temperature magnetic behavior. The magnetic elements of interest are manganese (Mn), cobalt (Co), iron (Fe), etc. Doping of ZnO NWs with one of the abovementioned elements will lead to the development of spintronics and other optomagnetic prototype devices [89,90]. In a series of synthesis experiments, it was possible to observe magnetic behavior of Co-doped ZnO NWs [82–84].

In this study the pure and Co-doped ZnO NWs with different Co (0%, 2%, and 5%) concentrations were synthesized as follows:

First, a seed layer of ZnO NPs 20–50 nm in size prepared as described before [65] were spun-coated on three different Sapphire substrate pieces of sizes. This ZnO NPs were spun-coated several times to insure that a uniform spatially distributed thin film has been fabricated. Then the samples were annealed at 120°C for 10 minutes. After this a 0.075 M HMTA solution was prepared and stirred for 1 hour. In a spate flask a Cobalt(II) acetate tetra-hydrate having different concentrations were prepared and added to the HMTA solution and stirred for 15 hours. A 0.075 M zinc nitrate hexahydrate was added dropwise to the above mixture and then the whole mixture was stirred for 3 hours [82]. Finally the Sapphire samples with the annealed seed layer where inserted vertically with the seed layer mounted upside down in the beaker containing the mixture and were placed in a laboratory over held at 90°C for 6 hours. The samples were then rinsed with deionized water several times and then blown with flowing nitrogen and left to dry at room temperature. Fig. 5.17 (top) shows two ZnO NWs samples grown on Sapphire substrate. These are pure ZnO NWs (sample S0) and 5% Co-doped ZnO NWs (5% Co). Both ZnO NWs shown in Fig. 5.17 (top) have a length of 3.5 μm, where their diameter was found to be 220 nm for S0, and 300 nm for S2. As can be seen both samples are formed of dense array of NWs that are vertically aligned. To investigate the incorporation of the Co inside the ZnO NWs are matrix electron paramagnetic resonance measurements were performed. Fig. 5.17 (bottom) shows the electron paramagnetic resonance (EPR) spectra of two samples S1 (2%) and S2 (5% Co) together with the simulated spectrum. The spectra shown in this figure is a clear indication of the fact that the ZnO NWs are single crystal. For $\theta = 0°$, the Co related signal displays a resolved eight lines structure, which is an indication of the incorporation of the Cobalt ions inside the ZnO crystal. These experiments are a clear indication of the fact that the low-temperature chemical synthesis can also be utilized for doping nanostructures with suitable elements. Indeed, this is a promising observation since doping is sometimes vital in modifying the tuning some desired properties.

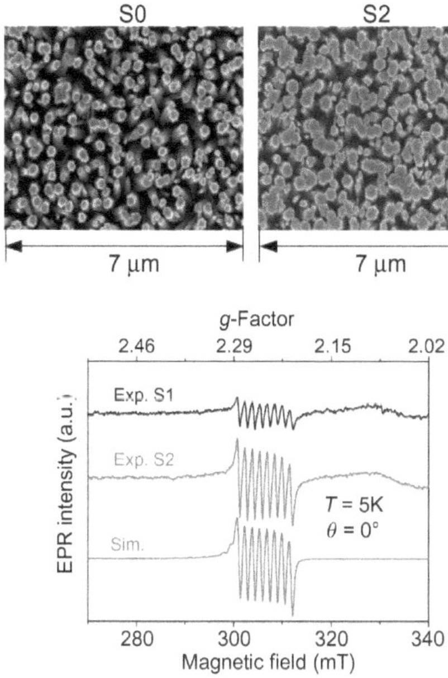

Figure 5.17 (Top) Two ZnO NWs samples S0 and S2 are pure, and 5% Co-doped ZnO NWs, respectively. (Bottom) Experimental electron paramagnetic resonance spectra of S1 (2% Co) and S2 (5%) doped ZnO NWs samples, recorded at 5K and for $\theta = 0$-degree orientation. The simulation of S2 spectrum is achieved using the parameters discussed in [82]. *NW*, Nanowire.

Although the low-temperature chemical synthesis of nanostructures has many advantages as discussed earlier, there are disadvantages too. One of these disadvantages is the low deposition rate. This is why to grow few micrometers of length; many hours are usually needed. To grow a typical ZnO NW of length of 2 μm, it takes about 6 hours, this is equivalent to a growth rate of about 5.5 nm/min. This can be an obstacle when considering the economic viability of producing large-scale nanostructures using the low-temperature chemical synthesis. Therefore, researchers have put their efforts to develop further the chemical approach to increase the synthesis rate. Such efforts will be discussed in the next section.

5.4 The electrochemical deposition

Electrochemical deposition is an aqueous chemical approach that provides some advantages over the conventional low-temperature aqueous chemical synthesis of metal oxide nanostructures [91]. Among these advantages is the increased deposition rate, which can be an obstacle when utilizing the conventional aqueous chemical growth. This technique was first used to deposit thin films of micro- and nanocrystals ranging in size from few micrometers down to few nanometers in size [91]. For coatings of sizes <10 nm deposited by this technique, superior properties were observed and consequently a wider interest of researchers has appeared [92]. When such a small grain seized nanocoating is deposited on a surface, the number of atoms/volume that are in contact with each other is then relatively very large compared to that of coatings of grains with much larger sizes. This can lead to much harder coating which is a target for durable materials. A typical electrochemical deposition cell for the synthesis of nanostructured metal oxide or any other material is shown in Fig. 5.18.

The typical electrochemical cell used for deposition/synthesis of nanostructures composed primarily of three electrodes connected to an electric current source. These are: the working electrode (where the deposition is intended), then a reference electrode, and finally a counter electrode (as shown in Fig. 5.18). The utilization of the electrochemical deposition/synthesis of nanomaterials can either be achieved by using a template free or templated substrate. As an example of a template free deposition/synthesis of nanostructure material, we consider the example

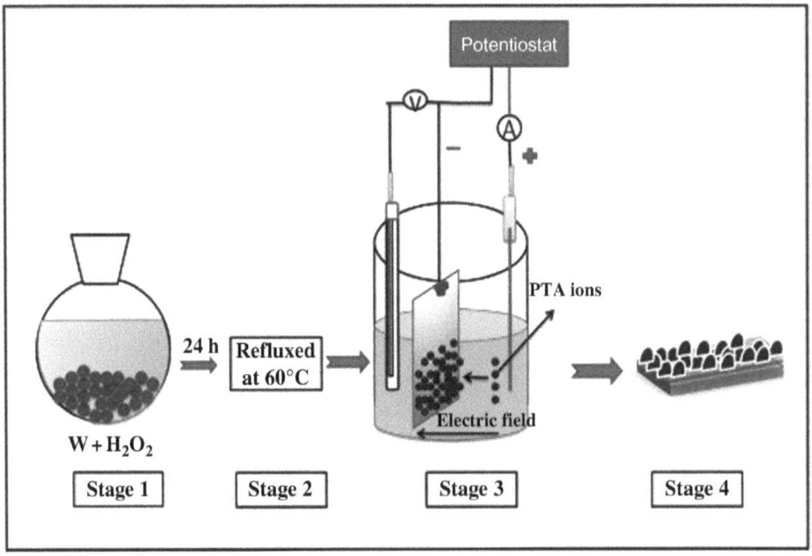

Figure 5.18 Schematic diagram showing the stages and the electrochemical cell used for the synthesis of WO_3 using a template free process [93].

of tungsten trioxide (WO_3) nanoflakes. Tungsten trioxide is one of the metal oxides that are of interest. It is n-type by unintentional growth, and it has an intriguing physiochemical property [93]. Electrochemical deposition/synthesis has been used to grow WO_3 of high quality [93]. WO_3 has been investigated as a potential candidate for many applications, for example, photocatalysis, gas sensors, thermoelectric devices, display devices, price labels, etc. For the synthesis of WO_3 analytical grade chemicals of high purity were used [93]. It is important to mention that for reproducible results in the synthesis of nanostructures analytical grade chemicals is a must, otherwise the synthesis cannot be reproduced.

We describe two different approaches, the first is achieved with a templated substrate while the second is when a free-templated substrate is used. For the first approach and to prepare the electrolyte for the electrodeposition of WO_3 peroxo tungstic acid was prepared. This preparation was performed by dissolving 1.25 g of tungsten powder into 40 mL 30% hydrogen peroxide (H_2O_2) and the mixture was kept under continuous stirring in a cold bath at a relatively low temperature range between 0°C and up to 10°C for a whole day (24 hours). After that and to remove the extra (H_2O_2) glacial acetic acid was added and the mixed solution was refluxed at 80°C for 6 hours under continuous stirring [94]. Finally, the peroxo tungstic acid was obtained by filtering. The peroxo tungstic acid was then diluted by adding water and isopropanol to reach a concentration of 50 mM of peroxo tungstic acid solution. The + purpose of the addition of the isopropanol is to increase the stability of the precursor solution by preventing the precipitation of the peroxo complex for relatively long period of time [94]. The final precursor solution was having a pH of 2.21. Then the pH is adjusted to three different values of 1.51, 0.9, and 0.3 to investigate the effect of the pH on the growth of the WO_3. The pH was adjusted by the addition of a suitable amount of oxalic acid. Hence, three different electrolyte solutions were used for this synthesis. To proceed with the synthesis of the WO_3, a $3 \times 1.5\ cm^{-2}$ florine tin oxide (FTO) coated glass substrates were used. At first, the FTO coated glass substrates were seeded by WO_3 NPs. Then a typical configuration of the electrochemical cell as that shown in Fig. 5.18 is used. In brief, using potentiometric conditions with an applied voltage of -0.45 V, the cell consisted of a platinum wire used as counter electrode, Ag/AgCl used as reference electrode and the FTO coated seed glass used as the working electrode. The synthesis was performed for 30 minutes (compared to at least 5 hours in case of the conventional chemical synthesis). When the synthesis duration was completed the FTO coated glass samples were removed and washed many times with deionized water. Finally, the samples were annealed at 400°C [94].

Fig. 5.19 shows SEM images of the three samples of WO_3 synthesized at different pH values. As can be seen the pH has a determinant role in

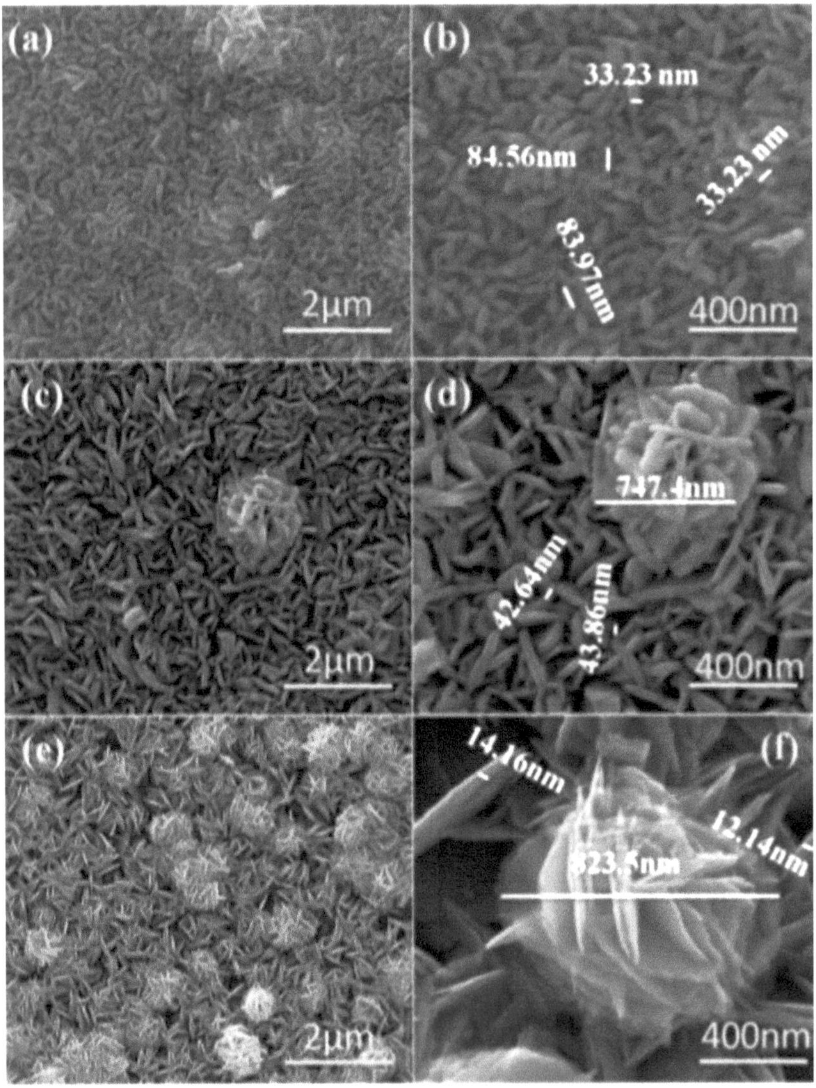

Figure 5.19 Scanning electron microscope of the WO_3 nanostructures synthesized by electrodeposition at (A) and (B) at a pH = 1.51, (C) and (D) at a pH = 0.9, and finally (E) and (F) at a pH = 0.3 [94].

the achieved morphology. For the case of a pH of 1.51 (Fig. 5.19A and B), the grown morphology observed is a thin film composed of small nanospindles of size range between 50 and 90 nm. While for the synthesis at pH = 0.9, nanoplatelets of larger thickness of about 40–50 nm is observed. When the pH was lowered to 0.3, the morphology has turned into homogenously distributed three-dimensional well-aligned hierarchical flower like structure with a total outer diameter of about 1 μm [94]. This hierarchical flower consists of nanoflakes having a thickness of about 20 nm. The variation of the pH in these synthesis experiments was achieved by adjusting the oxalic acid. Oxalic acid with its small legend molecules have a stabilizing effect and can easily bind with W^{6+} leading to stable tungstic acid during the WO_3 nucleation and synthesis [95]. Further research also indicated that the oxalic acid works as a surfactant of the formed WO_3 flakes and hind the coagulation of the flakes form forming a microstructure. Hence, the oxalic acid facilitates the formation of nanostructures over the formation of microstructures of WO_3 [96].

Fig. 5.20 illustrates schematically the morphological evolution of the WO_3 as the pH is varied. When the oxalic acid is added to the electrolyte solution, it will dissociate into conjugate base and hydronium ions. The effect of acid addition when WO_3 is synthesized by electrodeposition was reported to cause deflocculating due to the hydronium ions, and the conjugate base causes a separation of the ion complex that control the density and distribution of the WO_3 nuclei and affect their diffusion and grain formation tendency, and in addition it affects clustering of the solute prior the electrodeposition [97,98]. In these electrodeposition synthesis

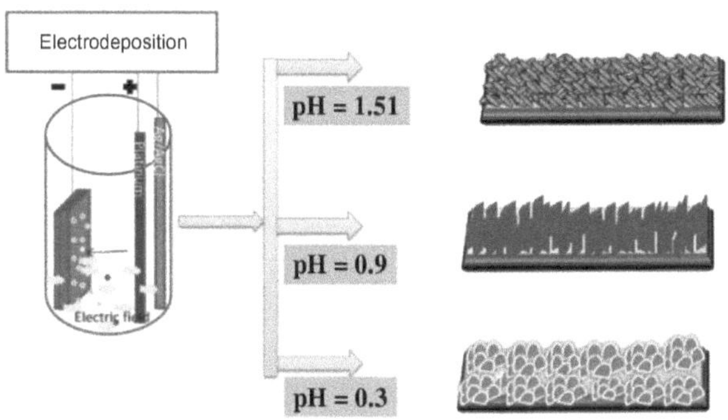

Figure 5.20 Schematic diagram showing the different morphologies achieved during electrodeposition of WO_3 due to the variation of the oxalic acid and hence variation of the pH of the electrolyte [94].

experiments for growing WO_3 nanostructures, the first pH value was 1.51 achieved by using 0.01 M oxalic acid concentration. At this pH value the synthesis has resulted in a nanoflakes like morphology as can be seen in Fig. 5.19A and B. By increasing the concentration of the oxalic acid to 0.05 M (pH = 0.3), this increase of the oxalic acid will lead to increase the hydronium ions leading to the growth of a well-defined hierarchical flower like morphology assembled by a thin (20–30 nm) WO_3 nanoflakes as can be seen in Fig. 5.19E and F. Hence, for the electrodeposition of WO_3, oxalic acid plays a vital role as a stabilizer, pH controller, and as a directing agent during the synthesis [94].

In another study, ZnO NRs were synthesized by both electrodeposition conventional chemical approach using the standard nutrients as before with the assistance of a seed layer [99]. Two different ZnO NRs were synthesized on GaN substrate. First both substrates were covered by a ZnO seed layer prepared as described earlier in this chapter. For the electrodeposition the electrolyte was prepared using equimolar 5 mM zinc nitrate hydrate ($Zn(NO_3)_2 \cdot 6H_2O$) and HMTA which was saturated with oxygen bubbles for 60 minutes before the electrodeposition [99]. The electrodeposition proceeded as usual by fixing the GaN substrate to a copper sheet and served as working electrode. A platinum wire is used as a counter electrode. The area of the GaN substrate was 1 cm^2 and the distance between the working and counter electrodes was about 3 cm. The ZnO NRs were synthesized catholically at a voltage of 1.1. V and a temperature of 90°C for a duration of 100 minutes [99]. While the other substrate was used to synthesize ZnO NRs using the standard chemical synthesis as described before at 90°C for a duration of 300 minutes [99]. Fig. 5.21 shows the ZnO NRs grown by the two methods. The thickness of the ZnO NRs was found to be 120–135 and 140–170 nm while the length was found to be 1.5 and 1.2 μm, for the electrodeposited and the chemically grown NRS, respectively [99]. These dimensions indicate that there is almost no difference between the two different ZnO NRs when using the electrodeposition or chemical synthesis approaches. However, the electrodeposited ZnO NRs required much less time to grow compared to those grown by the conventional chemical approach, that is, 100 versus 300 minutes. Further structural, electrical and optical characterization showed no significant difference between the two different ZnO NRs as will be elaborated below [99].

Fig. 5.22 shows a PL spectrum of the GaN substrate and the two other ZnO NRs samples grown by the electrodepositio0n and conventional

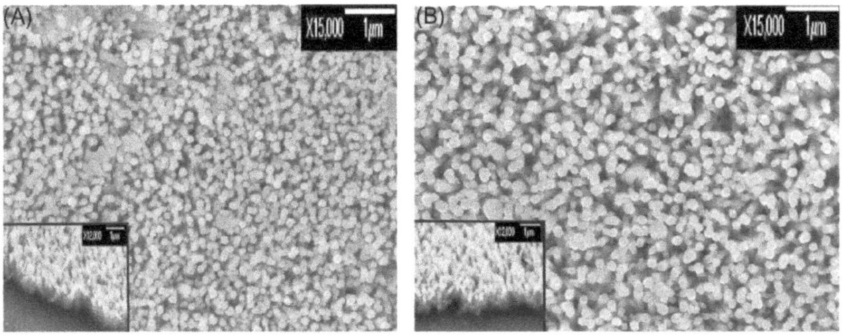

Figure 5.21 SEM images of ZnO NRs grown on *p*-GaN using (A) the electrodeposition technique, and (B) the aqueous chemical synthesis technique [99]. *NR*, Nanorod; *SEM*, scanning electron microscope.

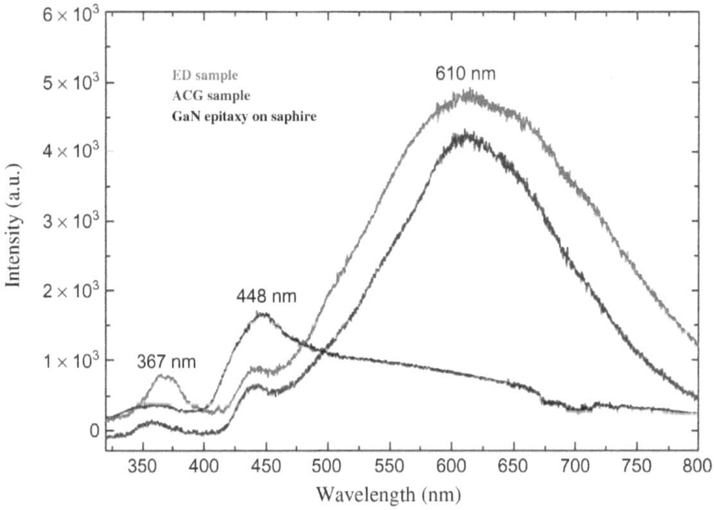

Figure 5.22 PL of the *p*-GaN substrate (*line with a peak at 448 nm*), the ZnO nanorods/*p*-GaN for the ED sample (*line with 367 nm peak*) and the conventional chemical synthesis sample (line with a peak below the line 610 nm) at room temperature using a Nd:YAG laser with an OPO and a frequency doubler and excitation wavelength at 280 nm [99]. *OPO*, Optical parameter oscillator; *PL*, photoluminescence.

chemical approaches. The general feature of the PL spectra in Fig. 5.22 is that it shows three different peaks in both the ZnO NRs grown samples. The first is the near band edge emission peak appearing at around 380 nm. This peak is an intrinsic property of wurtzite ZnO material. It is due to transition between the conduction and valence bands. The second peak at 446 nm is due to the magnesium doped GaN. While the third peak, the wide broad band centered at around 660 nm is due to recombination of carriers form the conduction band to oxygen interstitial combined with transition of Zn interstitial with holes trapped by oxygen interstitial [100–103]. Nevertheless, the PL of both ZnO NRs growing using the electroception and conventional chemical synthesis are almost identical in their morphology and other structural and optical properties. Hence, the main advantage of using the electrodeposition over the conventional chemical synthesis is the shorter time duration required for synthesis.

The solvothermal (see Section 5.8) chemical synthesis is another version of the chemical synthesis of nanostructures. The difference is that when utilizing the conventional chemical synthesis, the precursor solution is prepared using water, while for the solvothermal synthesis other solvents are used to dissolve specific salts [104]. The possibility of choosing different solvents when utilizing the solvothermal synthesis allows the synthesis of variety of materials including alloys, oxides sulfides, etc. It is obvious that the choice of the solvent plays a key role in the mechanism that lead to a specific material with specific morphology. In general, the solvothermal synthesis process can involve different reactions. These reactions can be classified to five different types. These are [104]: (1) oxidation–reduction, (2) hydrolysis, (3) thermolysis, (4) complex formation, and metathesis reactions. Due to the wide choice of the solvents, the solvothermal chemical approach has been utilized to synthesize a wide variety of nanostructures. Fig. 5.23 shows XRD pattern and SEM image of Cu_7Te_4

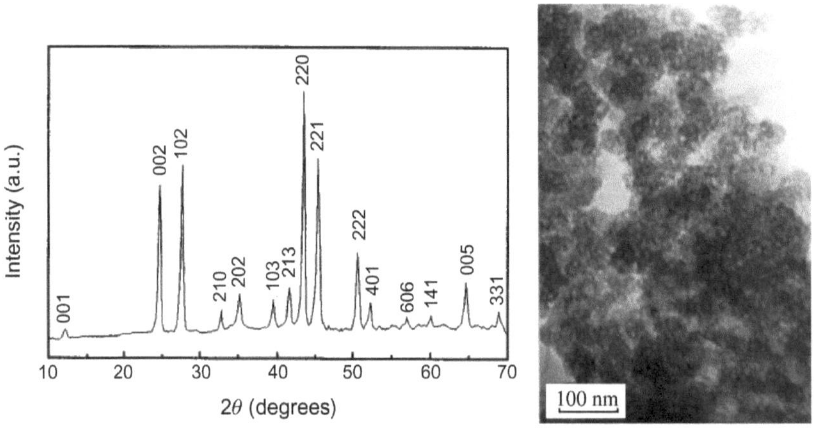

Figure 5.23 (left) X-ray diffraction pattern of the solvothermal as prepare nanocrystals of Cu_7Te_4, and (right) scanning electron microscope image of the same nanocrystal [105].

nanocrystals prepared using the solvothermal chemical method. The dominant reaction in a solvothermal chemical synthesis depends on the choice of the chemical and physical properties of the solvent. In a synthesis experiments, Cu_7Te_4 nanocrystals were grown by the solvothermal approach. Cu_7Te_4 belongs to the family of metal tellurides, which show metallic as well as covalent properties and of interest to application as a thermoelectric material. Fig. 5.23 shows the XRD peak and SEM image of the grown Cu_7Te_4 nanocrystals having an average particle size of 14 nm [105]. In these experiments, $CuCl_2$, H_2O, and tellurium were used as reagents while two different solvents were used separately. In one experiment, ethylenediamine was used as the solvent while in the second experiment, either diethylamine or benzene were used [105]. It was found that when diethylamine or benzene were used, tellurium did not react with $CuCl_2$. The difference between benzene and ethylenediamine is that the first is nonpolar while the second is a polarizing solvent and it will promote the solubility of the reagents and consequently it will facilitate the reaction leading to the synthesis of nanocrystalline Cu_7Te_4 [105]. In general, the solvothermal chemical synthesis has been used to synthesis a wide variety of nanostructured materials.

5.5 The microwave-assisted chemical deposition

There are many advantages of the conventional low-temperature (100°C) chemical synthesis route to demonstrate device quality nanomaterials, among them is easiness and no need of sophisticated instrumentation, the possibility to use soft substrate, good control of the emerging morphology, homogeneity, chemical composition, etc. Usually the chemical reaction takes place for a time period usually not less than 6 hours and it can extend to up to 48 hours. Such long growth duration time is a disadvantage. This relatively long synthesis duration is required when the conventional low-temperature aqueous chemical route is adopted can be an obstacle. As have been discussed before, the electrodeposition is another route that can reduce the synthesis duration by a factor of 3 or up to 5. Saying that time is important is valid for science. For synthetic chemistry definitely time is a very important factor because often many cycles of experiments are usually needed to develop a synthesis protocol for a new material and/or optimize the produced material regarding the quality, homogeneity, purity, yield, industrialization mass production cost, etc. Hence, any new synthesis routes that can reduce the time required to produce any new material would be very beneficial. Researchers have further developed an aqueous chemical route that can even reduce the synthesis time to a much higher degree. The so-called microwave-assisted chemical synthesis can in fact reduce the synthesis duration dramatically. This synthesis approach appeared for the first time in 1986 and was utilized for the synthesis of organic materials [106,107]. The advantages of microwave is that it can penetrate the material and at the same time supply energy and heat to all the volume of the material. Due to this microwave can lead to elevate the temperature of the reactants well above the boiling point of the synthesis nutrients. The possibility to elevate the temperature of the whole volume above the boiling point of the nutrients will lead to increase the synthesis speed by a factor of 10–1000 times [108,109]. Fig. 5.24 shows the difference in the temperature profile of conventional heating a microwave heating versus time over a period of 4 minutes.

Microwave radiation is the part of the electromagnetic radiation with a wavelength range lying between 1 and up to 1 mm. This is equivalent to a frequency range of 0.3–300 GHz. The use of microwave for the purpose of heating is based on the choice of the material to be heated. In connection to microwave heating the materials are divided into three different categories considering their interaction with microwaves [109]. The first are microwave reflectors, for example, metals. These are materials

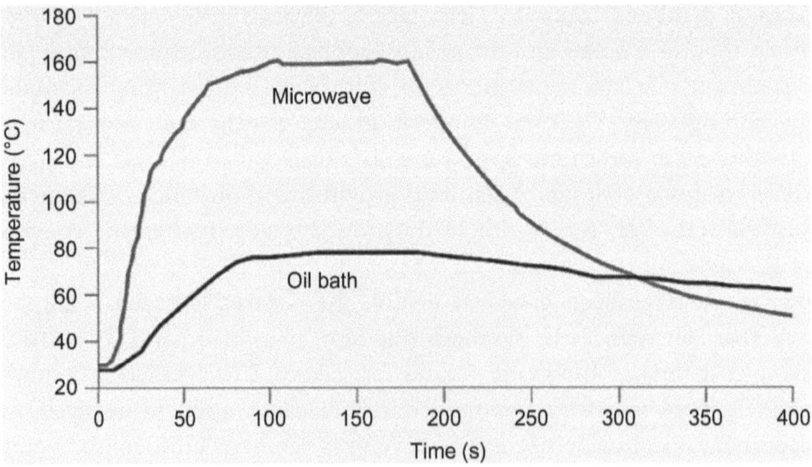

Figure 5.24 The temperature profile of a 5 mL ethanol sample subjected to conventional (temperature set at 100°C) and microwave heating (temperature set at 160°C) over 4 min [109].

that reflect microwave without any interaction. They are not affected by microwave heating. The second category are microwave transmitters, and these are materials that transmit microwave without absorbing them. Example of microwave transmitters is Teflon and quartz. Such materials transmit microwave without absorption of any heat generation, and therefore they are used as containers for other materials used for chemical synthesis of nanostructures. The third class of materials are those called microwave absorbers they are also called microwave high loss materials. These materials absorb the microwave radiation taking its energy and transfer it to heat energy very rapidly [109,110]. The heating process using microwave is a result of a dielectric loss. The electric part of the electromagnetic wave induces heating in microwave absorbers by two mechanisms. These two mechanisms are the ionic conduction and dipolar polarization [109]. When an electromagnetic filed is applied, the dipoles or the ion filed will start to align themselves with the alternating filed and at the same time energy is lost as heat due to molecules friction and dielectric loss. Hence in the material to be heated, if one species has a permanent dipole then the microwave irradiation will lose its energy as heat. Therefore, materials having polar molecules are microwave active, for example, ethanol, water, etc. Other nonpolar molecules, for example, toluene or benzene are microwave inactive [111].

The conversion of microwave active of the absorbed microwave energy at a specific frequency and temperature is governed by the so-called loss factor (tan σ) which is given as:

$$\tan \sigma = \frac{\varepsilon''}{\varepsilon'} \tag{5.6}$$

where is ε'' the dielectric loss, and ε' is the dielectric constant. The dielectric loss is a measure of the effectiveness of conversion of the microwave radiation into heat, while the dielectric constant is a measure of the polarizability in the presence of an electric field. A high value of tan σ is required and it means high absorption property of the microwave radiation in the specific medium and hence the ability of the medium to rapidly raise its temperature upon the exposure to microwave radiation [112].

The produced amounts of synthesized materials using the microwave-assisted chemical synthesis are usually characterized by high yield. Also, microwave-assisted chemical synthesis can be utilized for achieving nanostructures with relatively small size and at the same time having narrow size distribution with high purity and superior physiochemical properties [113].

Another important property characterizing microwave-assisted chemical synthesis is the fact that it is suitable for scaled-up processes for industry due to the advantage of the low thermal gradient effects [114]. Due to the interesting features of microwave-assisted chemical synthesis of nanostructures it has been used to synthesize many different morphologies of nanostructures both single and multimetal oxide materials. This technique has been utilized to synthesize both single and mixed metal oxides as will be shown below [109]. We here show some of examples of these demonstrated low-temperature microwave-assisted chemical synthesis experiments. As NPs are an important class of nanostructures, many different types of metallic and nonmetallic NPs have been synthesized by the microwave-assisted chemical method. All noble metals, including Ag, Au, and Pt NPs have been synthesized by the microwave-assisted chemical growth [115–117].

Fig. 5.25 shows a high-resolution TEM image of a part of a flower like nanostructure showing 1.8 nm size Au NPs arranged in a superstructure grown by the microwave-assisted chemical synthesis as will be described below. These Au small-size NPs were synthesized utilizing microwave induced route. Using an analytical grade chemical, the Au NPs shown in Fig. 5.25 were demonstrated. In brief, 0.2 mM $HAuCl_4$

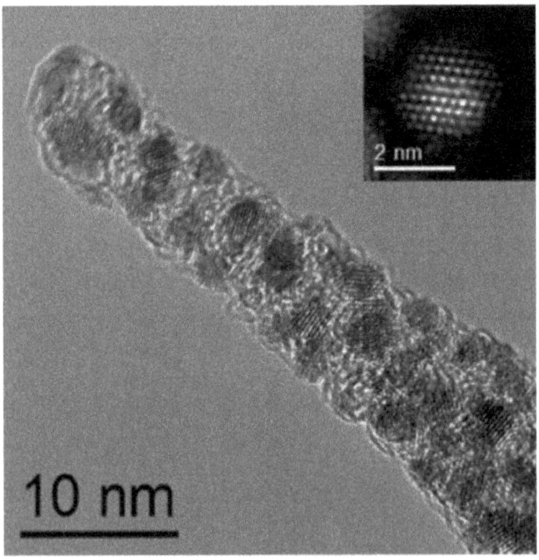

Figure 5.25 A high-resolution transmission electron microscope of a part of a flower like nanostructure showing Au nanoparticles arranged in a superstructure configuration. The insert shows a HAADF-STEM image of a single Au nanoparticle [115]. *HAADF-STEM*, High-angle annular dark-field scanning tunneling microscope.

was dissolved in 6 mL H_2O was then added to 1.25 mM of 1-dodecanethiol dissolved in 15 mL of toluene and the mixture was placed in a high-pressure vessel. Then the high-pressure vessel with the mixed chemicals was introduced to inside an ETHOS EZ Digestion System Micro Wave (Milestone, 2.5 GHz, sensor ATC400) equipment [115]. Then the system was irradiated with a power of 400 W for 60 seconds, followed by irradiation at 800 W for another 60 seconds, and this was followed by a final irradiation at 1200 W for 60 more seconds. The temperature of the vessel has then reached a temperature of 200°C. After the 3 minutes of microwave irradiation the synthesis was terminated, and the sample were left to cool down to room temperature and finally the NPs which appear as off-white powered was purified in ethanol [115]. As can be seen from Fig. 5.25 relatively small size (1.8 nm) Au NPs were obtained in a relatively very short synthesis duration of about 3 minutes. These Au NPs were passivated were observed to arrange themselves through a self-assembly route as self-supported superstructures of 1 mm diameter with an average thickness of about 400 nm [115]. X-ray results indicated that the *n*-alkanethiol molecules have interacted with the Au NPs and formed a cubic ordered array lying between the NPs and forming the superstructure as shown in Fig. 5.25 [115]. The superstructure of the Au NPs and the *n*-alkanethiol molecules has resulted in a gap between neighboring NPs with about 3.65 nm. Further characterization has indicated that no contaminants were found within the superstructures of Au NPs and the fact that this microwave-assisted chemical approach has led to excellent size and shape selectivity.

Other mixed oxide nanostructures have also been demonstrated using the microwave-assisted chemical synthesis. In fact, the microwave-assisted chemical synthesis route have been adopted to demonstrate a variety of binary and ternary metal oxide nanostructures. We here discuss an example of the microwave-assisted chemical synthesis of a ternary metal oxide nanostructure, namely Bismuth (III) vanadate ($BiVO_4$) [118]. $BiVO_4$ is a material that has been of interest for many technological applications. Among them is due to the observed absorption of the visible light radiation by this material, and consequently, the potential for photoprocesses realizing the sun visible radiation with all the sustainability benefits lying in such advantage. Bismuth vanadate exists in three different crystalline forms. These are tetragonal zircon, monoclinic scheelite, and tetragonal scheelite structure [118]. The band gap of the tetragonal form is reported to be 2.9 eV which falls within the UV absorption band, while the monoclinic $BiVO_4$ possesses a band gap of 2.4 eV which imply the possibility of visible light absorption band [118].

Using continuous stirring for 4 hours with heating of a mixture of a stoichiometric amount of V_2O_5 and NaOH in distilled water a solution of $NaVO_3$ was prepared and was used as a source of vanadium. Then another solution was prepared by dissolving Bi $(NO_3)_3 \cdot 5H_2O$ into concentrated HNO_3. Then for both these two solutions, CTAB was added while stirring for 10 minutes. After the 10 minutes stirring the two solutions were added and mixed with each other. It was observed that the mixed solution turned into an intensive orange-yellow color and the mixture solution pH was measured to be 1. After that the beaker containing the mixture was placed in a microwave oven and was irradiated for different durations ranging from 10 up to 40 minutes. These experiments were performed in a domestic microwave oven, however, this is not recommended, since nowadays there is a dedicated microwave oven for chemical synthesis of inorganic or organic nanocrystals (see below). The results of characterizing the $BiVO_4$ synthesized here using the microwave-assisted route indicated that it is possible to control the phase of the synthesized $BiVO_4$ by optimizing the microwave heating duration. It was shown that it is possible to prolong the microwave-assisted synthesis duration to obtain the desired monoclinic phase of the $BiVO_4$ [118].

More complex nanostructures have also been synthesized using the microwave-assisted chemical route. Antimony (Sb) doped Lead telluride/silver telluride ($PbTe/Ag_2Te$) core–shell composite nanostructures forming nanocubes have been synthesized using a microwave-assisted solvothermal chemical approach [119]. PbTe is an interesting narrow bandgap semiconductor having a value of 0.32 eV at room temperature and have a strong confinement effect within a large size due to the large average of the excitonic Bohr radius (~46 nm) [118]. Recent results have indicated that PbTe NWs are of interest as a thermoelectric material with high efficiency [120]. Further theoretical calculations have confirmed that the utilization of core–shell nanostructures composed of heterogenous materials can lead to improve the thermoelectric properties to a degree better than both materials by reducing the lattice thermal conductivity of the core–shell nanostructure to less than each of the two materials [119,121]. The microwave-assisted solvothermal synthesis of the $PbTe/Ag_2Te$ was performed in a microwave system designed for solvothermal synthesis (MDS-10, Sineo, Shanghai, China) [119]. This system is operated at 2.45 GHz using a power of a maximum of 1000 W. The internal temperature of the system is monitored using a high precision temperature and pressure sensors. For the synthesis a Teflon cylindrical shaped autoclave having a capacity of 100 mL was used. This vessel was placed inside the system at the idle and it was turning around under synthesis for obtaining a well distributed heat energy. The synthesis duration was 20 minutes and the temperature control accuracy was about ± 1°C. For the synthesis of the PbTe/Ag_2Te core–shell NWs, the following chemicals were purchased from Sinopharm Chemical Reagent Co., Ltd. and were used without further purification: antimony trichloride anhydrous ($SbCl_3$, 99.0%), silver nitrate ($AgNO_3$, 99.8%), lead acetate trihydrate (Pb $(CH_3COO)_2$–$3H_2O$, 99.5%), sodium tellurate (Na_2TeO_3, 98%), sodium borohydride ($NaBH_4$, 96%), sodium hydroxide (NaOH, 96%), ethylene glycol (EG, 99.9%) [119]. The synthesis proceeded as follows:

A 0.023 g $SbCl_3$, 0.017 g $AgNO_3$, 0.379 g $Pb(CH_3COO)_2 \cdot 3H_2O$, 0.266 g Na_2TeO_3, 1.2 g NaOH and 30 mL EG heated at 80°C were added to each other in a 100 mL vessel and were rigorously stirred at room temperature for 20 minutes. Then 0.2 g $NaBH_4$ was added to the mixture and then the vessel was stirred for extra 2 minutes. Then the Teflon autoclave was sealed, and microwave heated to 250°C and kept at this temperature for 20 minutes. After the 20 minutes the microwave heating was stopped, and the autoclave was left to cool down to room temperature naturally. After opening the autoclave, a precipitate was formed o and it was collected by centrifugation. Finally, the precipitate was washed thoroughly by deionized water and then by ethanol and was dried at 60°C in vacuum [118]. Fig. 5.26 shows some results of the characterization of the microwave-assisted solvothermal synthesized PbTe/Ag_2Te core/shell composite nanocubes [119]. These results indicated that using the microwave-assisted solvothermal chemical approach relatively rapid synthesis procedure has been demonstrated for such complex nanostructure core–shell NW with excellent result.

In many of the published papers on the microwave-assisted synthesis of nanostructures, the users have utilized domestic conventional microwave oven. However, this is not recommended at all, a specially dedicated microwave ovens are to be sued instead. This is because when using domestic microwave ovens, parameters like temperature, irradiation power, pressure, etc. are not monitored accurately, and hence the synthesis conditions are not determined accurately. To the contrary, modern dedicated microwave synthesis reactors are equipped with built-in magnetic stirrers, and sensors for monitoring the pressure and temperature, and hence the synthesis conditions can accurately be determined [122]. Moreover, in some of the microwave-assisted chemical synthesis processes,

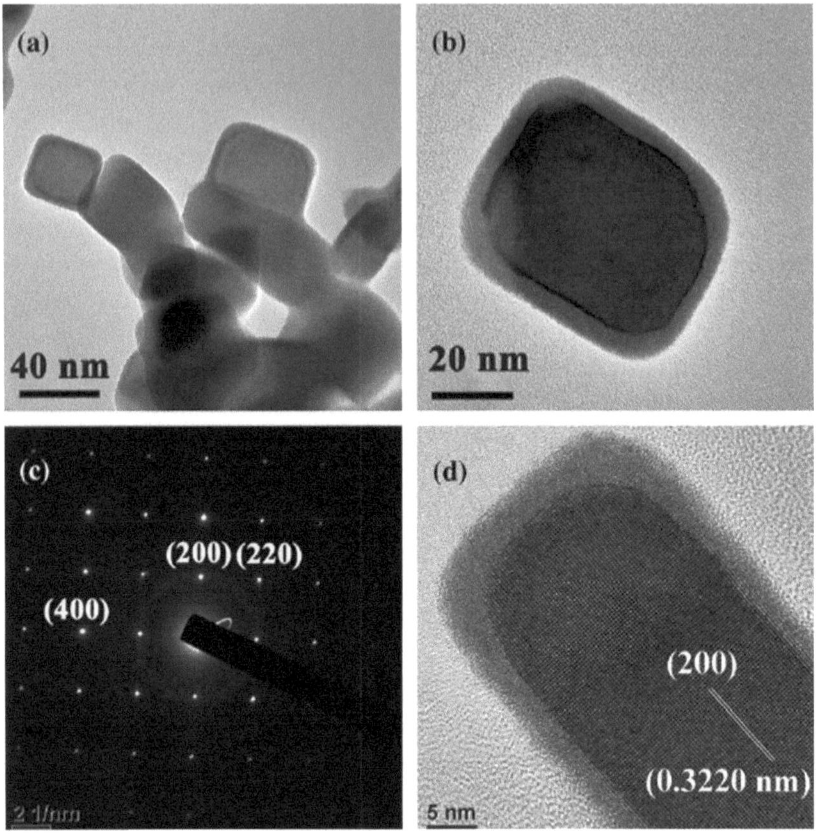

Figure 5.26 (A and B) TEM micrographs; (C) SAED pattern and finally in (D) HRTEM image of a single PbTe/Ag_2Te core/shell composite nanocubes [119]. *HRTEM*, High-resolution transmission electron microscope; *SAED*, Selected area electron diffraction; *TEM*, transmission electron microscope.

high absorption solvents with low boiling points are sued, and hence care must be taken. In most of the modern microwave-assisted synthesis reactors, a computer is connected to monitor all parameters, and hence a high security degree can be attained [122]. The utilization of the microwave-assisted reactor can in general be combined with all the different types of chemical-based aqueous synthesis approaches [122].

5.6 Solochemical synthesis of nanostructures

The solochemical process is another low-temperature aqueous chemical synthesis route suitable for the growth of high-quality and high-purity nanostructures. It is suitable for the synthesis of many metal oxide nanostructures. This method is based on the preparation of the precursor solution followed by a subsequent decomposition of the precursor at low temperature into the desired nanostructure [123]. An important feature of this process is that no by-products are produced and hence the chance of contamination is very low, also no templates, surface capping preparation, or catalysts are usually needed. Therefore, a high purity product can be obtained using the solochemical synthesis. This method can also be applied to prepare nanocomposites when choosing the correct precursors combination [124]. Below we present and discuss an example of the demonstration of a nanocomposite nanostructure using the solochemical method.

Zinc titanates are promising composite for dielectric applications. The $ZnO-TiO_2$ exists in three different compounds [124]. These are the Zn_2TiO_4 (cubic), $ZnTiO_3$ (hexagonal) and $Zn_2Ti_3O_8$ (cubic) [124]. Among these three compounds, the hexagonal $ZnTiO_3$ have superior electrical and dielectric constant compared to the other two compounds. The $ZnO-TiO_2$ was prepared by a two-step solochemical process as follows:

5.6.1 Zinc sol preparation

The zinc sol was prepared using concentrated NH_4OH with a 0.2 M $ZnCl_2$ to obtain $Zn(OH)_2$ precipitate. Then more NH_4OH was added to the precipitate was observed to be dissolved. The addition of the NH_4OH continued until a diluted Zn^{2+} complex sol was prepared.

5.6.2 Titanium sol preparation

Analytical grade titanium tetra isopropoxide (TTIP) (99.9%), citric acid, isopropanol (99.9%), and $C_6H_8O_7$ (99%) were bought from Merch and used without further purification. The v was dissolved together with polyhydroxy and the alcohol citric acid which is a chelating agent to produce TiO_2 powder. This was achieved as follows: using a nitrogen rich environment, the TTIP was dissolved in the isopropanol, then citric acid and deionized water were added followed by stirring for 2 hours. To prepare 0.1 M TiO_2 sol, a 0.05 mL of the citric acid, was added to 0.1 M of deionized water and 32 mL of isopropanol.

5.6.3 $ZnO-TiO_2$ nanocomposite solochemical synthesis

The two prepared sols were mixed together with a specific ratio and in order to obtain a nanocomposite with 50 mol.% Zn and 50 mol.% Ti and were then sonicated at 20 kHz for 5 minutes [124]. This process resulted in a milky solution. This mixture was then left to age for 3 days at room temperature. After aging the color changed to pale milky. Then by room temperature evaporation the pale milky powder was left to yield a dry powder. This dried powder was then subjected to calcination at different temperatures ranging from 500°C and up to 800°C for 1 hour [124]. The result of characterizing the produced and calcinated powder indicated that this two-step solochemical process has yielded a high-quality ZnO/TiO_2 nanocomposite material. As can be seen the approach is also quite simple with very few steps to yield the final product. It is also possible to scale-up the process in order to fabricate industrial quantities of such nanocomposite material.

5.7 Polyol chemical synthesis of nanostructures

The polyol chemical approach route is another process that was first developed for the synthesis of elemental metallic NPs [125]. The process is based on the reduction of an inorganic compound in the presence of high boiling point poly-alcohol (polyol). In some cases, the process is adjusted by adding water to inhibit the reduction reaction and consequently leading to the possibility of synthesizing different oxide NPs depending on the inorganic compound used [123]. When water is added, a forced hydrolysis accompanied by inorganic polymerization of the precursor inorganic sault will take place in the presence of the polyol [126,127]. When water is added to a polyol synthesis process to promote hydrolysis and polymerization, the formation of the oxides depends on the water/precursor molar ratio and on the reaction temperature [127,128]. The typical polyols utilized in this method are ethyl glycol, propylene glycol, and diethyl glycol [128]. One fundamental feature of this method is that the use of a polyol will lead to no agglomeration since the polyol will act as a protective agent. The morphological properties of the nanostructures synthesized using the polyol method can be manipulated by the different parameters, like temperature, precursor pH, the hydrolysis ratio, and the concentration of the polyol, etc. Although the polyol method was developed for the synthesis of metallic NPs, nowadays it is widely used strategy for the synthesis of a wide variety of nanostructured materials, like metals, oxides, chalcogenides, and nonmetallic elements [129]. In addition, polyols have a chelating property that provides an asset for controlling the NP size with possible control of the agglomeration [129]. In addition, and for the synthesis of nanomaterials, the adjustment of the synthesis parameters to achieve precise phase with high purity is a challenging issue. The polyol approach is usually used to obtain exact phase with high purity of binary metal oxide nanostructures. However, the method has proven, when precisely adjusting the synthesis parameters, to be successful in obtaining precise ternary high purity metal oxide nanostructures of a specific phase composition [129]. Here we show an example of the synthesis of ternary compounds of transition metal tungsten oxide nanocompounds [MWO_4 (M = Mn, Fe, Co, Ni, Cu, Zn)] using the polyol approach at low temperature (<100°C) [130].

The family of ternary tungsten oxide is an inorganic functional material of interest for many applications. They are often prepared by precipitation methods; however, the products are usually amorphous. Of interest in this wide family is the divalent metal tungstate (MWO_4) [130]. These divalent ternary tungstates are usually formed in two crystalline types. These are the scheelite-type and the wolframite-type [130]. The first is formed for larger cations, while the second is formed for small cations. Atypical difficulty when synthesizing ternary compounds is the formation of binary oxides, that is, phase segregation. In addition to this the reduction of metallic cations present in the synthesis medium must be avoided. These two unwanted effects are not easy to avoid if synthesizing the ternary compounds by precipitation approach.

In order to achieve a high purity single phase ternary compound nanomaterial synthesis without the formation of different binary oxide phases (phase segregation) and without metal reduction, all synthesis parameters including the starting synthesis precursors, synthesis temperature, duration, pH, etc. have all to be carefully chosen. This careful choice of the synthesis parameters and its effect on the final produced ternary compound can be illustrated when synthesizing $CuWO_4$ [130]. When using diethylene glycol (DEG) together with either copper halides or Cu $(NO_3)_2 \cdot 5H_2O$ at a temperature lower than 100°C, two different final products are observed. For the case of using copper halides, elemental copper is formed even at such low temperature because of the high electrochemical potential of copper. However, when using $Cu(NO_3)_2 \cdot 5H_2O$ as the starting precursor material, an oxidative effect due to the nitrate anion will be present, the formation of metallic copper can be hindered. However, the chosen temperature for the synthesis can affect the final produced material. If a temperature higher than 200°C is chosen, then a mixture of binary oxide, for example, Cu_2O and CuO can be formed. But when the synthesis temperature is lower to less than 200°C, a ternary

$CuWO_4 \cdot 2H_2O$ is formed [130]. Also adjusting the pH to an optimum value is vital because it will affect the formation of binary tungsten oxide if the value of it is too low.

The polyol-mediated chemical synthesis of the MWO_4 NPs (M = Mn, Fe, Co, Ni, Cu, Zn) transition metal ternary oxide was achieved using low cost metal saults like metal chlorides to reduce the cost. The polyol used in these experiments was DEG under controlled pH condition to avoid unwanted effects as discussed above. To control the pH an acetate buffer solution was prepared using acetate precursor with addition of hydrochloric acid. A solution of $M(CH_3COO)_2(M(NO_3)_2$ in the case of M = Mn, Fe, Co, Ni, Cu, or Zn) was prepared and was added to 25 mL of DEG at a temperature of 70°C and a pH value of 4.8–5.5. Then this solution was added to 25 mL equimolar solution of $Na_2WO_4 \cdot 2H_2O$. The addition of the last solution was followed by immediate nucleation of the ternary corresponding compound NP according to the M used where M = Mn, Fe, Co, Ni, Cu, or Zn. Then the temperature was raised to 220°C for 1 hour. During this last stage, the water was distilled out of the synthesis mixture. Then the remaining solution with the suspended NPs was left to cool down to room temperature naturally. The centrifugation was applied to separate the NPs form the solution. Then the obtained powder was resuspended into ethanol and glacial acetic acid, and then centrifugation was applied once more. By doing so all the extra DEG as well as other residual saults were completely removed, and a pure ternary compound powder was obtained [130]. Fig. 5.27 shows the full series of the wolframite-type transition metal tungstates MWO_4 (M = Mn, Fe, Co, Ni, Cu, and Zn) that were obtained using the polyol-mediated chemical synthesis approach. Viewing the SEM of Fig. 5.27 indicate that the NPs are uniform with sphere like or lens like morphology with a narrow size distribution. By studying 100 NPs, the size distribution as found to be between 3 and 12 nm. While the specific surface area of the powdered samples was found to lie between 15 and 212 m^2/g [130]. These experiments indicate that the polyol approach is a power full approach as it can in a one-step process provide pure ternary transition metal NP with excellent size distribution and with absence of any binary oxide or metallic phase.

5.8 Solvothermal chemical synthesis

Solvothermal reaction is generally defined as "a chemical reaction taking place in a closed system in the presence of a solvent at a temperature

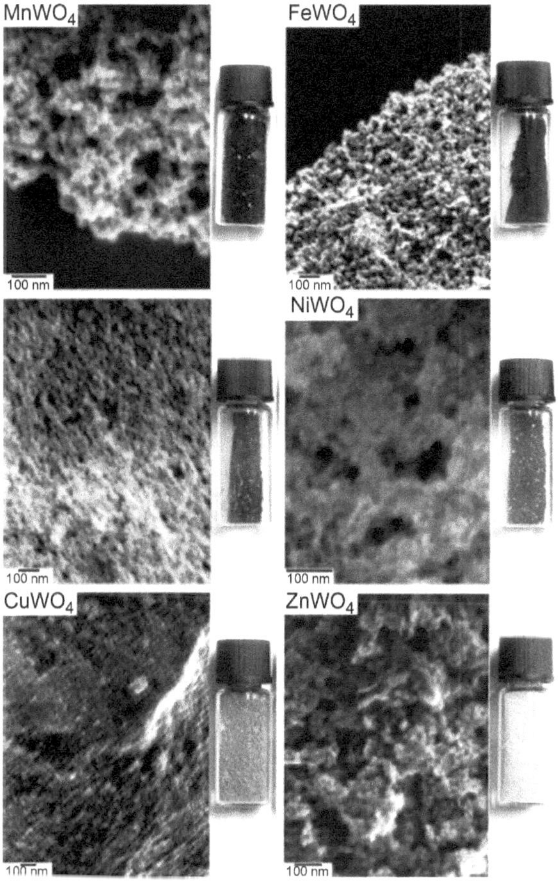

Figure 5.27 Scanning electron microscope images and micrographs of different ternary transitional metal tungsten nanoparticles MWO_4 (M = Mn, Fe, Co, Ni, Cu, and Zn) synthesized using the polyol approach [130].

higher than the boiling point of the solvent" [104]. The solvent can either be aqueous (water based) or nonaqueous. When water is used as the solvent, the reaction is called hydrothermal. Sometimes, a mixture of water and another nonaqueous solvent are utilized, and in this case the reaction is called solvohydrothermal reaction. With this definition of the solvothermal reaction, it is obvious that all the above different types of chemical approaches are in a way a "solvothermal approach" since they all utilize solvents. Because the solvothermal reaction takes place in a closed system, the pressure during the reaction is relatively high. There are two different main parameters governing the solvothermal reactions. These are chemical parameters and thermodynamically parameters [104]. The chemical parameters include the type of solvent, type of precursors sued, and the mixing approach. While the thermodynamically parameters includes, the temperature and pressure. Among these the chemical parameters, mainly the precursor concentration also enters as an internal parameter that can affect the morphology of the resulted nanomaterial [131]. In general, chemical synthesis approaches based on solvothermal processes being aqueous and performed at high pressure and mild temperature or non-aqueous are promising for material science and specially for nanomaterials utilization.

5.9 Successive ionic-layer adsorption and reaction

In 1985 Nicolau presented a chemical low-temperature process called successive ionic-layer adsorption and reaction (SILAR) [132]. Although this technique was introduced as a technique for the synthesis of solid compound thin films, nowadays the SILAR technique is used for the synthesis of complex nanostructures, for example, core–shell structures for example.

The SILAR reaction is based on the adsorption reaction of ions from solutions with rinsing by deionized water between different immersions steps [132,133]. The purpose of this rinsing is to avoid homogenous precipitation of the required material to be grown on a solid surface. A building block of the SILAR approach is hence based on the collection of a material on another solid material surface. As adsorption is defined as an interfacial layer existing between two different phases of a system, the SILAR approach is an adsorption-based method. In general, when two heterogeneous systems are brought together absorption is expected to take place. This means that adsorption can take place for three different systems. These are solid–gas, liquid–solid or gas–liquid systems [133]. The SILAR process happens when we have a specific the liquid–solid system adsorption reaction. Adsorption is an exothermic process taking place at the surface of a substrate in the presence of ions. What makes the SILAR process possible is the attraction force of the ions in the solution and the surface of the solid material. The forces of attraction leading to the SILAR process can be of different nature. The force of attraction can be chemical attractive forces, or cohesive forces or van de Walls force [133]. These forces exist because ions in solutions carry charge and the atoms at the surface of the solid have dangling bonds that lead to charge the surface molecule. There are many factors affecting the adsorption process leading to the SILAR process. These are the system temperature, pressure, the substrate nature, the molar concentration of the solution, and the area of the solid substrate surface exposed to the ionic solution. In the case of complex morphologies, the SILAR method is usually combined with another synthesized nanostructure most probably by using another approach. We here show one example of combing the SILAR approach with the hydrothermal synthesis to obtain CuS/ZnO heterostructure.

ZnO nanostructures are among the most researched materials in recent years. This is owing to the many excellent properties of this material combined with the possibility of synthesis using many chemical low-temperature approaches. Among the different application of ZnO is the photocatalysis applications. Although ZnO can be used under visible light excitation due to the mid-bandgap centers, it is usually utilized under UV due to its relatively large band gap [134]. Researchers have modified ZnO properties by doping with different metals to increase the visible light based photocatalytic activity of ZnO [135–137]. To fully utilize ZnO different nanostructure morphologies for photocatalysis under visible light, other approaches have been published. Among them is the combination of ZnO used as a carrier/support for another nanostructure material have been suggested [134]. The aim will be to extend the photoresponse of the ZnO and increase the efficiency under visible light irradiation and to reach this aim, wide bandgap material should be combined with a narrow bandgap material. As ZnO is a nontoxic and can be synthesized with low cost, the narrow bandgap material to be combined with ZnO is preferably to possess the same properties. A suitable candidate is cupper sulfide (CuS). Below we briefly present a CuS/ZnO heterostructure that has been achieved by synthesizing CuS NPs using the SILAR method on ZnO NWs synthesized by the hydrothermal low-temperature chemical approach.

In the experiment to synthesize ZnO NWs decorated by CuS NPs, first, ZnO was synthesized on a stainless-steel mesh substrate. The ZnO NWs were synthesized at 70°C using the hydrothermal chemical approach discussed above, and after that the SILAR method was applied to decorate the ZnO NWs with CuS NPs as will be described below. To obtain the CuS NPs the su8bstrate containing the ZnO NWs was immersed into a Na_2S solution (50 mM) and followed by immersion into a $CuSo_4$ (25 mM) aqueous solution. Then the sample was thoroughly washed by deionized water and was blown by nitrogen for drying. This immersion and rinsing sequence were repeated different times (from 0 for bare ZnO NWs and up to 10 times to obtain CuS decorated ZnO NWs) to obtain different samples. It is to be noted that the SILAR immersion cycles have been performed at room temperature. Fig. 5.28 (left) shows some characterization results, while Fig. 5.28 (right) displays a schematic diagram showing the decorated NW and the process of activation under visible light irradiation [134]. Fig. 5.29 shows the UV–visible absorption spectra of CuO/ZnO decorated NWs subjected to different SILAR cycles. AS can be clearly seen the absorption of the visible radiation is increasing with the increase of the SILAR cycle [134].

5.10 Summary of the low-temperature chemical synthesis of nanostructures

In this chapter the most common and well-established low-temperature chemical synthesis methods have been presented. Although there are many other uncommon low-temperature chemical methods not discussed, the presented method clearly shows that the presented methods have successfully yielded device quality material, as will be further highlighted in the next chapter.

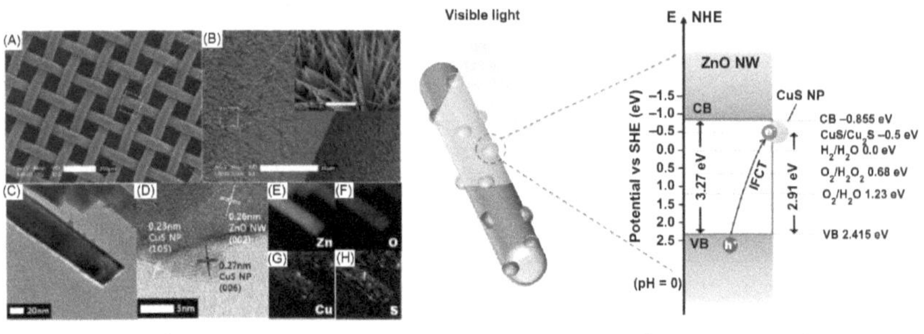

Figure 5.28 (Left) Scanning electron microscope images of (A) CuS/ZnO heterostructure nanowires grown on the stainless-steel mesh substrate and (B) low magnification of the top-view image. The inset is a high-magnification top-view image of (B). Transmission electron microscope images of CuS/ZnO heterostructure nanowires: (C) a low-magnification image and (D) a high-resolution transmission electron microscope image. (E–H) EELS elemental mapping images of the Zn, O, Cu, and S, respectively. (Right) Schematic illustration for the CuS/ZnO heterostructure nanowires showing the visible light activated proposed interfacial charge transfer process from the valence band of ZnO nanowires to the CuS nanoparticles [134]. *EELS*, Electron energy loss spectroscopy.

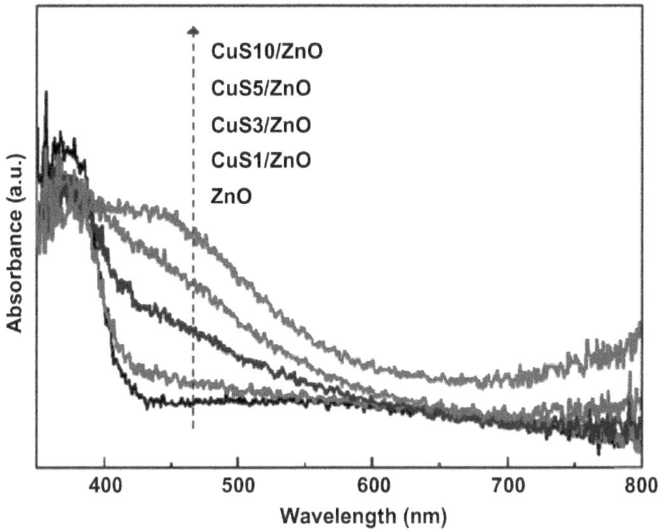

Figure 5.29 Different spectra of CuS/ZnO decorated nanowires achieved by different SILAR cycles for the deposition of the CuS nanoparticles [134]. SILAR, Successive ionic-layer adsorption and reaction.

References

[1] G.M. Whitesides, J.P. Mathias, C.T. Seto, Science 254 (1991) 1312.
[2] S. Mathur, H. Shen, S. Barth, C. Cavelius, SPIE Newsroom. <http://spie.org/newsroom/0080-inorganic-nanomaterials-through-chemical-design?ArticleID = x8864&SSO = 1>.
[3] S. Mathur, H. Shen, Inorganic nanomaterials from molecular templates, in: H.S. Nalwa (Ed.), Encyclopedia of Nanoscience and Nanotechnology, vol. 4, American Scientific Publisher, 2004, p. 131.
[4] G.M. Whitesides, B. Grzybowski, Science 295 (2002) 2418.
[5] T. Trindade, P. O'Brien, N.L. Pickett, Chem. Mater. 13 (2001) 3843.
[6] Y.N. Xia, P.D. Yang, Y.G. Sun, Y.Y. Wu, B. Mayers, B. Gates, et al., Adv. Mater. 15 (2003) 353.
[7] S. Mathur, H. Shen, in: H.S. Nalwa (Ed.), Encyclopedia of Nanoscience and Nanotechnology, vol. 4, American Scientific Publisher, 2004.
[8] E. Matjevic, Acc. Chem. Res. 14 (1981) 22.
[9] D. Sinclair, K.V. LaMer, Chem. Rev. 245 (1949) 44.
[10] K.V. LaMer, M.D. Barnes, J. Colloid Sci. 1 (1946) 71.
[11] I. Johnson, V.K. LaMer, J. Am. Chem. Soc. 69 (1946) 1184.
[12] V.K. LaMer, R.H. Dinegar, J. Am. Chem. Soc. 72 (1950) 4847.
[13] V.K. LaMer, Ind. Eng. Chem. 44 (1952) 1270.
[14] N. Rajput, Int. J. Adv. Eng. Tech. 7 (2015) 1806.
[15] L. Ayuk Eugene, O. Ugwu Marigoretti, B. Aromnimo Samuel, Chem. Res. J. 2 (2017) 97.
[16] W. Ostwald, Chemie 22 (1897) 289.
[17] L.L. Schramm, E.N. Stasiuk, D.G. Marangoni, Annu. Rep. Prog. Chem. Sect. C 99 (2003) 3.

[18] Y.B. Molla Mahaleh, S.K. Sadrnezhaad, D. Hosseini, J. Nanomater. 2008 (2008). ID 47095.
[19] C.B. Murray, C.R. Kagan, M.G. Bawendi, Annu. Rev. Mater. Sci. 30 (2000) 545.
[20] J.Th. G. Overbeek, Adv. Colloid Interface Sci. 15 (1982) 77.
[21] C.B. Murray, D.J. Norris, M.G. Bawendi, J. Am. Chem. Soc. 115 (1993) 8706.
[22] M.L. Steigerwald, A.P. Alvisatos, J.M. Gibson, T.D. Harris, R. Kortan, A.J. Muller, et al., J. Am. Chem. Soc. 110 (1987) 3046.
[23] M.R. Detty, M.D. Seildler, J. Org. Chem. 47 (1982) 1354.
[24] R.A. Zingaro, B.H. Steeves, K.J. Irgolic, Organomet. Chem. 4 (1965) 320.
[25] R.C. Mbwahnche, L.B. Matyushkin, O.A. Ryzhov, O.A. Aleksandrova, V.A. Moshnikov, J. Phys. Conf. Ser. 741 (2016) 012026.
[26] W. Ehrfeld, Ind. Eng. Chem. Res. 38 (1999) 1075.
[27] K.F. Jensen, Chem. Eng. Sci. 56 (2001) 293.
[28] M. Green, P. O'Brien, Chem. Commun. 1912 (2001).
[29] H. Nakamura, Y. Yamaguchi, M. Miyazaki, H. Maeda, M. Ueharab, P. Mulvaneyc, Chem. Commun. 1844 (2002).
[30] V. Mantzaris Nikos, Chem. Eng. Sci. 60 (2005) 4749.
[31] E. Curti, Co-precipitation of radionuclides: basic concepts, literature review and first applications, INIS 29 (1997) 4.
[32] O.G. Grendal, A.B. Blichfeld, S.L. Skjærvø, W. van Beek, S.M. Selbach, T. Grande, et al., Crystals 8 (2018) 253.
[33] M. Avrami, J. Chem. Phys. 7 (1939) 1103.
[34] M. Avrami, J. Chem. Phys. 8 (1940) 212.
[35] J.R. Eltzholtz, C. Tyrsted, K.M.O. Jensen, M. Bremholm, M. Christensen, J. Becker-Christensen, et al., Nanoscale 5 (2013) 2372.
[36] P. Nørby, M. Roelsgaard, M. Søndergaard, B.B. Iversen, Cryst. Growth Des. 16 (2016) 834.
[37] H.L. Andersen, K.M.Ø. Jensen, C. Tyrsted, E.D. Bøjesen, M. Christensen, Cryst. Growth Des. 14 (2014) 1307.
[38] J.D. Hancock, J.H. Sharp, J. Am. Ceram. Soc. 55 (1972) 74.
[39] J. Becker, M. Bremholm, C. Tyrsted, B. Pauw, K.M.Ø. Jensen, J. Eltzholt, et al., J. Appl. Crystallogr. 43 (2010) 729.
[40] S. Cheong, J. Watt, B. Ingham, M.F. Toney, R.D. Tilley, J. Am. Chem. Soc. 131 (2009) 14592.
[41] M.J. Whitaker, R.D. Bayliss, F.J. Berry, C.J. Greaves, Mater. Chem. 21 (2011) 14523.
[42] G. Westin, M.J. Nygren, Mater. Sci. 27 (1992) 1617.
[43] A.M. Abakumov, M.G. Rozova, E.V. Antipov, J. Hadermann, G. Van Tendeloo, M.V. Lobanov, et al., Chem. Mater. 17 (2005) 1123.
[44] Z.W. Pan, Z.R. Dai, Z.L. Wang, Science 291 (2001) 1947.
[45] M. Willander, O. Nur, Q.X. Zhao, L.L. Yang, M. Lorenz, B.Q. Cao, et al., Nanotechnology 20 (2009) 332001.
[46] M.L. Fuller, J. Appl. Phys. 15 (1944) 164.
[47] M. Willander, O. Nur, N. Bano, K. Sultana, New J. Phys. 11 (2009) 125020.
[48] M. Willander, L.L. Yang, A. Wadeasa, S.U. Ali, M.H. Asif, Q.X. Zhao, et al., J. Mater. Chem. 19 (2009) 1008.
[49] N. Bano, S. Zaman, A. Zainelabdin, S. Hussain, I. Hussain, O. Nur, et al., J. Appl. Phys. 108 (2010) 043103.
[50] M.H. Asif, S.M. Usman Ali, O. Nur, M. Willander, U.H. Englund, F. Elinder, Biosens. Bioelectron. 26 (2010) 1118.
[51] L. Vayssieres, A. Hagfeldt, S.-E. Lindquist, Pure Appl. Chem. 72 (2000).
[52] C.N.R. Rao, Pure Appl. Chem 69 (1997) 199.

[53] C.N.R. Rao, Pure Appl. Chem. 66 (1994) 1765.
[54] L. Vayssieres, Ph.D. thesis, Université Pierre et Marie Curie, Paris, 1995.
[55] L. Vayssieres, N. Beermann, S.-E. Lindquist, A. Hagfeldt, Chem. Mater. 13 (2001) 233.
[56] L. Vayssieres, K. Keis, S.-E. Lindquist, A. Hagfeldt, J. Phys. Chem. 105 (2001) 3350.
[57] L. Vayssieres, Adv. Mater. 15 (2003) 464.
[58] R.C. Markle, Nanotechnology 11 (2000) 89.
[59] J.C. Garcia, J.F. Justo, W.V.M. Machado, L.V.C. Assali, Phys. Rev. B. 80 (2009) 125421.
[60] R. Parize, J. Garnier, O. Chaix-Pluchery, C. Verrier, E. Appert, V. Consonni, J. Phys. Chem. C 120 (2016) 5242.
[61] L. Schmidt-Mende, L. MacManus-Driscoll, Mater. Today 10 (2007) 40.
[62] M. Andrés Vergés, A. Mifsud, C.J. Serna, J. Chem. Soc. Faraday Trans. 86 (1990) 959.
[63] L.E. Greene, M. Law, J. Goldberger, F. Kim, J.C. Johnson, Y. Zhang, et al., Angew. Chem. Int. Ed. 42 (2003) 3031.
[64] L.E. Greene, M. Law, D.H. Tan, M. Montano, J. Goldberger, G. Somorjai, et al., Nano. Lett. 5 (2003) 1231.
[65] C. Pacholski, A. Kornowski, H. Weller, Angew. Chem. 41 (2002) 1188.
[66] Z. Chen, L. Gao, J. Cryst. Growth 293 (2006) 522.
[67] Z. Wang, X.F. Qian, J. Yin, Z.K. Zhu, Langmuir 20 (2004) 3441.
[68] D.S. Boyle, K. Govender, P. O'Brien, Chem. Commun. 0 (2002) 80.
[69] H. Zhang, J. Feng, J. Wang, M. Zhang, Mater. Lett. 59 (2005) 169.
[70] S. Elhag, Z.H. Ibupoto, V. Khranovskyy, M. Willander, O. Nur, Vacuum 116 (2016) 21.
[71] U. Pal, P. Santiago, J. Phys. Chem. B 109 (2005) 15317.
[72] B.P. Pichon, C. Leuvrey, D. Ihiawakrim, D. Tichit, C. Gerardin, J. Phys. Chem. C 115 (2011) 23695.
[73] H. Alnoor, C.O. Chey, G. Pozina, X. Liu, V. Khranovskyy, M. Willander, et al., AIP Adv. 5 (2015) 087180.
[74] H. Alnoor, G. Pozina, X. Liu, V. Khranovskyy, D. Iandolo, M. Willander, et al., J. Appl. Phys. 119 (2016) 165702.
[75] H. Alnoor, G. Pozina, M. Willander, O. Nur, Phys. Status Solidi A 21 (2017) 1600333.
[76] A. Zainelabdin, S. Zaman, G. Amin, O. Nur, M. Willander, J. Cryst. Growth 10 (2010) 3250.
[77] A.S. Myerson, second ed., Handbook of Industrial Crystallization, 45, Butterworth-Heinemann, Boston, MA, 2001.
[78] L.L. Yang, Q.X. Zhao, M. Willander, J. Alloys Compd. 469 (2009) 623.
[79] M. Guo, P. Diao, S.J. Cai, Solid State Chem. 178 (2005) 1864.
[80] C.B. Tay, H.Q. Le, S.J. Chua, K.P.J. Loh, Electrochem. Soc. K45 (2007) 154.
[81] Z. Zhou, Y.J. Deng, Phys. Chem. C 113 (2009) 19853.
[82] A. Savoyant, H. Alnoor, S. Bertaina, M. Willander, O. Nur, Nanotechnology 28 (2017) 035705.
[83] H. Alnoor, A. Savoyant, X. Liu, G. Pozina, M. Willander, O. Nur, J. Appl. Phys. 121 (2017) 215102.
[84] A. Savoyant, H. Alnoor, O. Pilone, O. Nur, M. Willander, Nanotechnology 28 (2017) 285705.
[85] R. Shabannia, J. Mol. Struct. 1118 (2016) 157.
[86] A.K. Rana, P. Bankar, Y. Kumar, M.A. More, D.J. Late, P.M. Shirage, RSC Adv. 6 (2016) 104318.
[87] B. Panigrahy, M. Aslam, D. Bahadur, J. Phys. Chem. C 114 (2010) 11758.

[88] O. Lupan, L. Chow, L.K. Ono, B.R. Cuenya, G. Chai, H. Khallaf, et al., J. Phys. Chem. C 114 (2010) 12401.
[89] K.R. Kittilstved, W.K. Liu, D.R. Gamelin, Nat. Mater. 5 (2006) 291.
[90] Z.H. Zhang, X. Wang, J.B. Xu, S. Muller, C. Ronning, Q. Li, Nat. Nanotechnol. 4 (2009) 523.
[91] M. Ulzaki, T. Omi, J. Electrochem. Soc. 143 (1996) L53.
[92] I. Gurrappa, L. Binder, Sci. Technol. Adv. Mater. 9 (2008) 043001.
[93] S. Poongodi, P.S. Kumar, D. Mangalaraj, N. Ponpandian, P. Meena, Y. Masuda, et al., J. Alloys Compd. 719 (2017) 71.
[94] S. Poongodi, P.S. Kumar, Y. Masuda, D. Mangalaraj, N. Ponpandian, C. Viswanathan, et al., RSC Adv. 5 (2015) 96416.
[95] S.H. Baeck, T. Jaramillo, G.D. Stucky, E.W. McFarland, Nano. Lett. 2 (2002) 831.
[96] M.U. Sun, X.U. Ning, Y.W. Cao, J.N. Yao, E.G. Wang, J. Mater. Sci. Lett. 19 (2000) 1407.
[97] W.L. Kwong, A. Nakaruk, P. Koshy, C.C. Sorrell, J. Phys. Chem. C 117 (2013) 17766.
[98] W.L. Kwong, A. Nakaruk, P. Koshy, C.C. Sorrell, Thin Solid Films 154 (2013) 191.
[99] S. Kishwar, K. ul Hasan, N.H. Alvi, P. Klason, O. Nur, M. Willander, Superlattices Microstruct. 49 (2011) 32.
[100] E. Lai, W. Kim, P. Yang, J. Nano Res. 1 (2008) 123.
[101] C.H. Ahn, Y.Y. Kim, D.C. Kim, S.K. Mohanta, H.K. Cho, J. Appl. Phys. 105 (2009) 013502.
[102] M. Ghosh, A.K. Raychaudhuri, Nanotechnology 19 (2008) 445704.
[103] M. Willander, O. Nur, J.R. Sadaf, M.Q. Israr, S. Zaman, A. Zainelabdin, et al., Materials 3 (2010) 2643.
[104] G. Demazeau, J. Mater. Sci. 43 (2008) 2104.
[105] B. Li, Y. Xie, J.X. Huang, H.L. Su, Y.T. Qian, J. Solid State Chem. 146 (1999) 47.
[106] R.J. Giguere, T.L. Bray, S.M. Duncan, G. Majetich, Tetrahedron Lett. 27 (1986) 4945.
[107] R. Gedye, F. Smith, K. Westaway, H. Ali, L. Baldisera, L. Laberge, et al., Tetrahedron Lett. 27 (1986) 279.
[108] B.A. Roberts, C.R. Strauss, Acc. Chem. Res. 38 (2005) 653.
[109] A. Mizraei, G. Neri, Sens. Actuators, B: Chem. 237 (2016) 749.
[110] K.J. Rao, B. Vaidyanathan, M. Ganguli, P.A. Ramakrishnan, Chem. Mater. 11 (1999) 882.
[111] A. Lew, P.O. Krutzik, M.E. Hart, A.R. Chamberlin, J. Comb. Chem. 4 (2002) 95.
[112] J.P. Tierney, P. Lidstrom, Microwave Assisted Organic Synthesis, Blackwell Publishing Ltd, Australia, 2005.
[113] S.C. Motshekga, S.K. Pillai, S.S. Ray, K. Jalama, R.W.M. Krause, J. Nanomater. 12 (2012) 1.
[114] A.B. Panda, G. Glaspell, M.S. El-Shall, J. Am. Chem. Soc. 128 (2006) 2790.
[115] C. Gutiérrez-Wing, R. Esparza, C. Vargas-Hernández, M.E. Fernández García, M. José-Yacamán, Nanoscale 4 (2012) 2281.
[116] S. Komarneni, D. Li, B. Newalkar, H. Katsuki, A.S. Bhalla, Langmuir 18 (2002) 5959.
[117] M.N. Nadagouda, T.F. Speth, R.S. Varma, Acc. Chem. Res. 44 (2010) 469.
[118] H.M. Zhang, B.J. Liu, H. Wang, W.X. Zhang, H. Yan, J. Nanopart. Res. 10 (2008) 767.
[119] G.H. Dong, Y.J. Zhu, Chem. Eng. J. 193–194 (2012) 227.
[120] G. Tai, B. Zhou, W.L. Guo, Phys. Chem. C 112 (2008) 11314.

[121] R.G. Yang, G. Chen, M.S. Dresselhaus, Nano Lett. 5 (2005) 1111.
[122] I. Bilecka, M. Niederberger, Nanoscale 2 (2010) 1269.
[123] M. Gusatti, C.E.M. Campos, J.A. Rosário, D.A.R. Souza, N.C. Kuhnen, H.G. Riella, J. Nanosci. Nanotechnol. 11 (2011) 5187.
[124] M.R. Vaezi, J. Mater. Process. Technol. 205 (2008) 332.
[125] F. Fievet, J.P. Lagier, B. Blin, B. Beaudoin, M. Figlarz, Solid State Ionics 32/33 (1989) 198.
[126] C. Feldman, H.O. Jungk, Angew. Chem. 113 (2001) 372.
[127] C. Feldmann, Adv. Funct. Mater. 13 (2003) 101.
[128] F. Fievet, R. Brayner, The polyol process, in: R. Brayner, F. Fievet, T. Coradin (Eds.), Nanomaterials: A Danger or a Promise? - A Chemical and Biological Perspective, Springer, London, 2013, pp. 1–26.
[129] H. Dong, Y.-C. Chen, C. Feldmann, Green Chem. 17 (2015) 4107.
[130] J. Ungelenk, M. Speldrich, R. Dronskowski, C. Feldmann, Solid State Sci. 31 (2014) 62.
[131] Q. Wang, D. Pan, S. Jiang, X. Ji, L. An, B. Jiang, J. Cryst. Growth 286 (2006) 83.
[132] Y.F. Nicolau, Appl. Surf. Sci. 22/23 (1985) 1061.
[133] H.M. Pathyan, C.D. Lokhande, Bull. Mater. Sci. 27 (2004) 85.
[134] M. Lee, K. Yong, Nanotechnology 23 (2012) 194014.
[135] J.X. Wang, X.W. Sun, Y. Yang, H. Huang, Y.C. Lee, O.K. Tan, et al., Nanotechnology 17 (2006) 4995.
[136] J. Song, S. Lim, J. Phys. Chem. C 111 (2007) 596.
[137] S.H. Yi, S.K. Choi, J.M. Jang, J.A. Kim, W.G. Jung, J. Colloid Interface Sci. 313 (2007) 705.

Further reading

Weinberg, 1987 M.C. Weinberg, J. Cryst. Growth 82 (1987) 779.

CHAPTER 6

Emerging new applications

6.1 Introduction

It is of no doubt that the most important discovery of our modern era is the discovery of the transistor reported on the 23rd of December 1947 at Bell Laboratories in Murray Hill, New Jersey. The discovery of the transistor led to the replacement of the relatively larger size cathode ray tubes and valves used to control electronic signals before the existence of the transistor. The progress of modern electronics has progressed quickly after the demonstration of the transistor. Without the transistor life today would probably look like it was 100 years ago. The discovery of the transistor has opened the way for the development of large-scale or even very large-scale integrated circuits (ICs). Today electronic processors based on transistors contains hundreds of millions of transistors integrated and connected to each other in an area of a few square centimeters. The main drawback of ICs is their high production cost. Traditionally ICs electronics production lines have manufactured such electronic circuits by relying on high-temperature processes steps, and using mainly silicon (Si) as the material for production. These high-temperature processing steps hence require that such processes be performed in well controlled clean environments, that is, highly controlled clean rooms. This is necessary in order to avoid contaminants that can stick and contaminate the production environment forever when subjected to high temperature. In addition, supporting substrates that carry such high-temperature produced electronic circuits should also be resistant to changes under such relatively elevated temperatures. Most of the electronic ICs available today are based on Si. Although Si's properties are not the best among the different semiconductors we know today, it has dominated due to many reasons. Among these reasons is its abundant availability and also the possibility to grow high-quality insulating oxide using Si. One of the main drawbacks of Si is its poor optical properties.

As has been discussed in the previous chapters, the low-temperature synthesis of nanomaterials has opened a new possibility for obtaining device quality materials at the nanoscale at temperatures well below 100°C.

Low Temperature Chemical Nanofabrication
DOI: https://doi.org/10.1016/B978-0-12-813345-3.00006-X

Due to the relatively small footprint of these nanostructures, they do not require lattice mismatched substrate to carry them. This allows for the possibility of integrating nanomaterials with excellent properties on Si ICs. Such a possibility will broaden the scope of the application of monolithic Si ICs to the optoelectronic applications. In addition to this, nanomaterials, due to their outstanding properties, have led to the emergence of many other new applications. The absorption and emission of different small-size nanoscale materials that cover a relatively wide range from high-energy ultraviolet down to infrared radiation have led to the emergence of many interesting applications. Another important aspect of a synthesis temperature well below 100°C is the possibility of using nonstandard and flexible substrates. Such a possibility will have a positive impact on the development of flexible electronics, for example, displays. There have been many other interesting emerging applications as a consequence of the low-temperature chemical synthesis of nanomaterials. As nanostructures have a small "footprint," they can nucleate and grow on such substrates without the need for a lattice matching property. This adds another advantage to the flexible substrates, that is, they do not have to be of the same lattice structure of the overgrown nanomaterial. Therefore, many nanomaterials of high-crystalline quality grown on nonconventional substrates, for example glass, plastic, paper, and textile, have been utilized for different applications [1–4]. Usually such nanomaterials grown on such nonconventional substrates can be highly flexible, with extra advantageous properties, like being resistive to fatigue strain, and hence they are suitable for complying with the requirements of future electronic systems [5]. In the past polymers were thought to be the best alternative for such flexible electronics, however they are sensitive to fatigue strain and, in addition, inorganic nanomaterials are superior in their semiconducting properties.

In this last chapter we will present some of such interesting emerging applications. The focus will be on the most demonstrated applications for both technical and biomedical research areas.

6.2 Emerging sensors

As has been discussed in Chapter 2, Phenomenon at the nanoscale, one of the direct consequences of reducing the size of materials to the nanoscale is the increase of the sensitivity due to the relatively high increase of the surface area to volume ratio. Clearly this is of direct positive impact on sensors in general. Therefore, one of the most researched areas and published papers using nanomaterials are on sensors in general. In addition, the utilization of relatively low-temperature chemical synthesis of nanostructure has added another advantage, that is, the use of flexible substrates. Using the low-temperature chemical approach for synthesizing nanomaterials, many different types of sensors were demonstrated. We will here present and discuss some of these sensors.

Among the different classes of sensors, self-powered sensors are of potential due to their suitability for operating in remote areas for longer periods of time. Using the low-temperature chemical synthesis, we will focus the discussion on two innovative class of self-powered sensors, these are the intracellular potentiometric sensors and metal detection colorimetric sensors. The low-temperature chemical synthesis allows soft and fragile materials to be utilized as substrates for the overgrowth of nanomaterials. The overgrowth of ZnO nanorods on a thin fragile microcapillary glass as a substrate was demonstrated as described in Ref. [6,7].

To develop selective intracellular potentiometric sensors, a borosilicate glass capillary having an inner diameter of about 0.5 μm and an outer diameter of about 07 μm with a total length of about 50 mm was used [6]. The first step was to coat the glass capillary with a thin silver layer using conventional physical evaporation. To prepare a silver/sliver chloride (Ag/AgCl) reference electrode, one silver-coated capillary was dipped into 0.2 M hydrochloric acid and electrolyzation was applied using a voltage of 1 V for one minute. Using another silver-coated capillary glass tip, ZnO nanorods were grown using the low-temperature synthesis approach described in the previous chapter. The ZnO nanorods were only grown on a portion of the tip of the capillary, different portions with lengths between 10 μm and up to 3 mm were prepared. Fig. 6.1 displays scanning electron microscope images at different magnifications of the ZnO nanorods overgrown on the tip of the borosilicate glass capillary [6]. As can be seen, ZnO nanorods with a hexagonal face perpendicularly were grown with a diameter ranging between 100 and 120 nm and a length of about 900–1000 nm. Dibenzo-18-crown-6, when incorporated into polyvinylchloride (PVC) membranes, behaves as an ionophore-based ion selective electrode for calcium (Ca^{2+}) ions. To obtain an ion selective electrode, the electrode must be functionalized with a specific material. Dibenzo-18-crown-6 when added to a PVC will act as a Ca^{2+} selective electrode. To functionalize the presented intracellular ion selective electrode, powdered PVC is dissolved into 5 mL tetrahydrofuran with the

Figure 6.1 A typical SEM image of the ZnO nanorods grown on Ag-coated capillary using low-temperature growth, before PVC coating. The figure shows a ZnO nanorods–coated tip at different magnifications [6]. *SEM*, Scanning electron microscope.

addition of 10 mg of dibutyl phthalate to act as a plasticizer and 10 mg of $Ca2^{+}$ specific ionophore (DB18C6) [6]. After the preparation of the selective material the capillary thin glass tube with the ZnO nanorods grown on top was dipped into the Ca^{2+} selectivity solution and left to dry at room temperature. Using the prepared intracellular ion selective electrode, external calibration was performed versus an Ag/AgCl reference electrode by using a $CaCl_2$ solution having a dynamic concentration range of 100 nM to 10 mM.

After calibrating the working electrode, Ca^{2+} was selectively measured intracellularly for two types of cells. The two types of cells used were human adipocytes (fat cells) and frog oocytes (egg cells). Using a micromanipulation probe the two electrodes, reference and working electrodes, were carefully inserted inside the cell under investigation. Fig. 6.2 shows human adipocytes under measurement [6]. The human adipocytes were isolated by collagenase digestion from pieces of adipose tissue during surgery at the University hospital at Linköping University, Sweden. The frog cells were handled by following the method approved by the local animal care committee at Linköping University, Sweden. The pair of electrodes that were used for the intracellular ion selective measurements were mounted and manipulated as described above and shown in Fig. 6.2. In such a configuration the electromotive force generated by ions around the working electrode surface is then measured by a sensitive voltmeter. The amount of measured electromotive cell voltage will be proportional to the amount of the ions present. This is achieved by using the calibration curve values. The measurement of the free calcium ions in both cells was

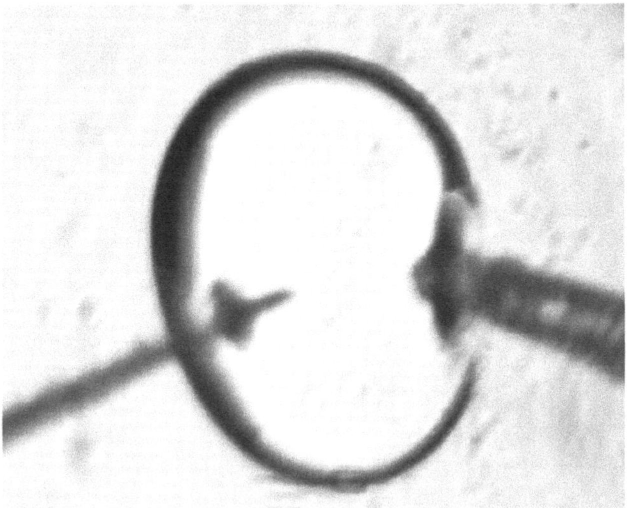

Figure 6.2 A typical microscope image of a single adipocytes human fat cell during measurement [6].

found to be 123 ± 23 nM for the human adipocytes and for the frog egg cells the measured calcium ion concentration was found to be 250 ± 50 nM [6]. These values are consistent and very close to values of the free calcium ions measured in the same cells using other methods reported in the literature [8,9]. To verify the measurements and the fact that the calcium ions detected do belong to the concentration inside the cells, that is, intracellular concentration, two different configurations were tested. These two configurations are shown in Fig. 6.3.

As can be seen in Fig. 6.3, two different working electrodes were used to measure the intracellular free calcium ion concentration. For the case shown in Fig. 6.3A, the working electrode used was having functionalized ZnO nanorods outside the cell, while in the case shown in Fig. 6.3B, all the functionalized ZnO nanorods were inserted inside the cell under measurement. Then the calcium ion concentration was varied in the buffer solution in both cases. Only for the case shown in Fig. 6.3A was the calcium ion concentration observed to vary [6]. This is an indication and verification that the measured calcium ion concentration given above for the two different types of cells do in fact represent the intracellular free calcium ion concentration. It is clear that the low-temperature chemical synthesis has allowed the utilization of fragile submicrometer glass capillaries for such direct intracellular measurements. Indeed a deep knowledge

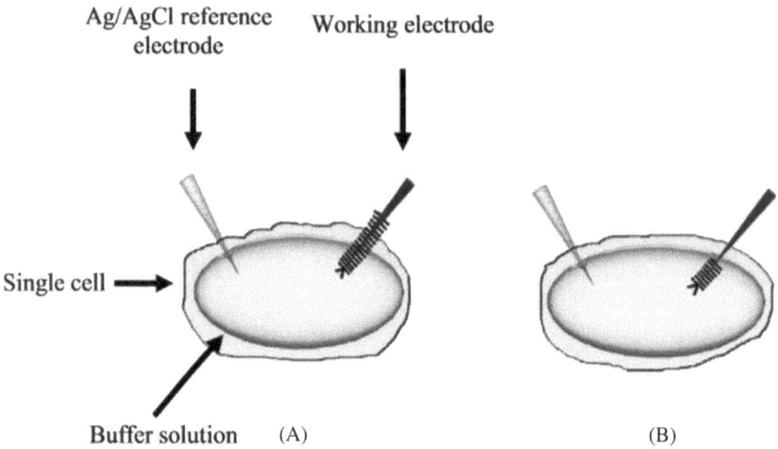

Figure 6.3 Schematic diagram displaying the two different configurations used to verify the intracellular measurements. In (A) part of the functionalized ZnO nanorods are in contact with the buffer layer, while in (B) all the functionalized ZnO nanorods are inserted inside the cell [6].

using direct methods to understand the physics and chemistry of the multipath processes in cells is one of the dreams of scientists, as such processes are only investigated using indirect methods of analysis. Such nanostructures deposited on the surface of submicrometer glass capillary can be used for obtaining such knowledge [10–12].

The second example of a sensor prepared using the low-temperature chemical approach to be presented here is a disposable colorimetric sensor for the detection of metals in aqueous solutions [13–15]. The utilization of the benefit of zinc oxide material, which has been a very popular semiconducting material in nanoscience due to its many interesting properties for a variety of applications, has also been extended by forming composites in different configurations with other materials. Among the interesting configurations that have been demonstrated using ZnO is core–shell structures. As has been discussed in an early chapter in this book, core–shell structures can provide properties different from both composite materials and can provide a wide variety of band gap engineering leading to the new property mentioned above. Among the interesting core–shell structures are core–shell nanoparticles (NPs). Recently zinc oxide–zinc sulfide core–shell NPs (ZnO@ZnS) have been demonstrated using a variety of methods. However, by using the coprecipitation low-temperature chemical synthesis with two simple steps, a simple, fast, and reliable method has been reported for the synthesis of ZnO@ZnS NPs [13,14].

The synthesis of this core–shell structure starts with the formation of the ZnO core NPs. For synthesizing the ZnO core NPs, aqueous solutions of zinc acetate dihydrate (0.5 M) and sodium hydroxide (1 M) were prepared separately. Then by using two different pipettes the two aqueous solutions were poured into a beaker at room temperature and the mixture was stirred for 2 hours until the transparent mixture turned into a whitish milky mixture. Then the resulted precipitate was separated by centrifugation and was washed by deionized water and acetone. Finally, the clean precipitate was dried in a standard clean laboratory oven at 75°C [13]. The synthesis of this core–shell structure starts with the formation of the ZnO core NPs. For synthesizing the ZnO core NPs, aqueous solutions of zinc acetate dehydrate (0.5 M) and sodium hydroxide (1 M) were prepared separately. Then by using two different pipettes the two aqueous solutions were poured into a beaker at room temperature and the mixture was stirred for 2 hours until the transparent mixture turned into a whitish milky mixture. Then the resulted precipitate was separated by centrifugation and was washed by deionized water and acetone. Finally, the clean precipitate was dried in a standard clean laboratory oven at 75°C [13]. To over cover the ZnO NPs with ZnS and form the core–shell NP 0.3 g of the ZnO NPs were sonicated into 50 mL isopropanol for 5 minutes, then by using NaOH the pH of the solution was adjusted to 10. Then a solution of Na_2S was dropwise added to the ZnO NPs solution and it was subjected to continuous stirring for 1 hour at 60°C. This was followed by continuous stirring for 1 hour [13]. Then finally a 0.05 M solution of $ZnCl_2$ was dropwise added to the above mixture and the new mixture was stirred for 1 hour. The molar concentration of the Na_2S added at first was varied (M = 0.025, 0.05, and 0.075) to optimize the synthesis of the ZnO@ZnS core–shell NPs [13]. Fig. 6.4 shows the schematic diagram of the synthesis of the ZnO@ZnS core–shell NPs [13].

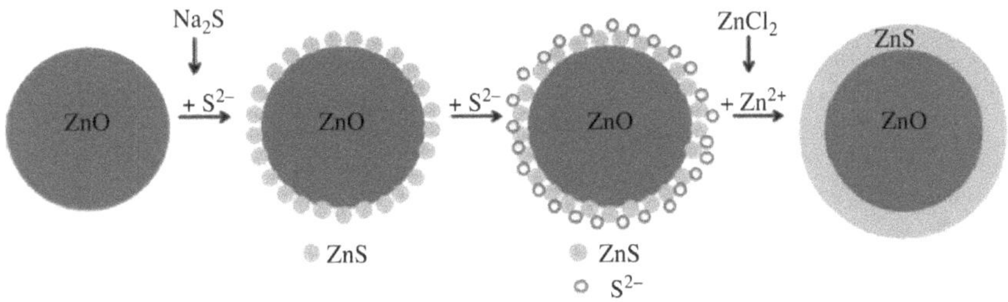

Figure 6.4 A schematic illustration of the ZnO@ZnS CSNPs formation mechanism [13]. *CSNPs*, core-shell nano-particles.

Fig. 6.5 shows different transmission electron microscope images of a ZnO@ZnS core–shell NP grown as described above. The grown ZnO@ZnS core–shell NPs were used to prepare a colorimetric self-powered sensor for the detection of ionic metals in aqueous solutions. To prepare such a sensor on a paper substrate, 0.5% (W/V) solution was prepared form the as-grown ZnO@ZnS core–shell NPs. This solution was subjected to ultrasonication for 5 minutes followed by 30 minutes of stirring [15]. Then a small amount (10 μL) of this solution was slowly dropped onto the surface of a normal piece of paper and was kept at room temperature for 30 minutes. Then different standard solutions of cupper ions (Cu^{2+}) having 15, 75, 150,

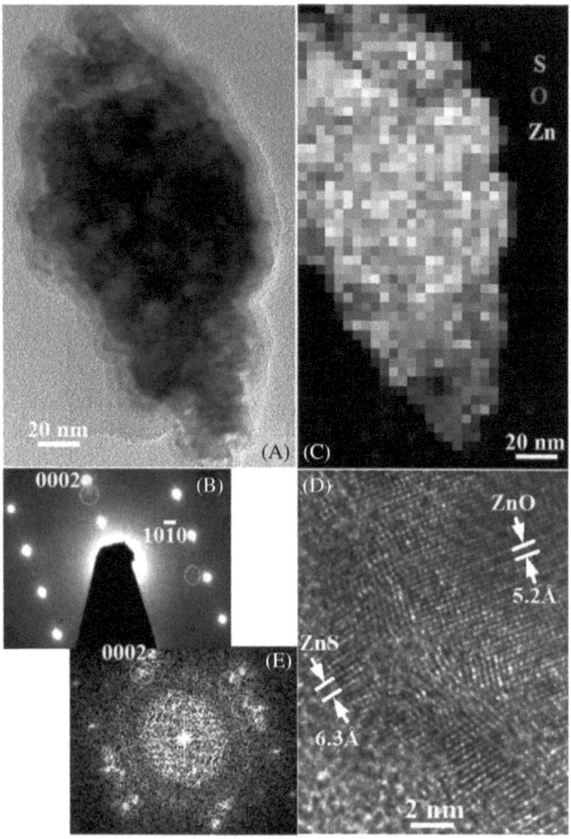

Figure 6.5 (A) Bright field transmission electron microscope image of a ZnO@ZnS core–shell nanoparticle, (B) corresponding SAED pattern from (A). (C) EDX mapping exhibiting S, O, and Zn distribution, (D) high-resolution transmission electron microscope image from the rectangle region on the left of (A), and (E) FFT pattern from (D) [13]. *EDX*, Energy dispersive x-ray spectroscopy; *SAED*, Selected area diffraction.

300, 750, and 1500 μM concentrations were prepared using deionized water (18 MΩ resistivity) and used for the testing of the colorimetric self-powered sensor. The paper sensor was inserted into the solutions having the specific known concentrations of the Cu^{2+}. Then after few minutes the different paper sensors were photographed using a 13 Megapixel camera. Then the images were transferred to a computer to measure the color intensity. Using a special software, the color change of the paper sensor was related to the amount of Cu^{2+} in the tested aqueous solution.

Fig. 6.6A shows the color intensity versus the concentration of the Cu^{2+} obtained by imaging the paper colorimetric sensor, while in Fig. 6.6B the same color intensity is plotted with its numerical value

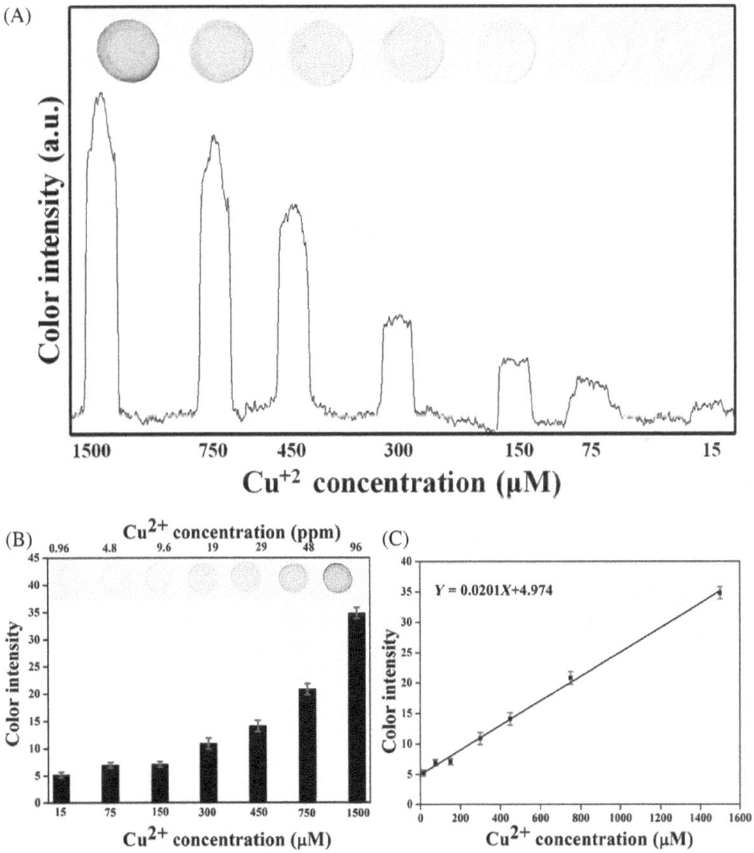

Figure 6.6 (A) Color intensity versus the Cu^{2+} ions concentration obtained from the image of the paper using a digital photograph as an inset, (B) the value of the color intensity versus the Cu^{2+} ions concentration, and (C) the calibration curve of the color intensity at different concentrations of the Cu^{2+} ions [15].

versus the Cu^{2+} concentration. The mechanism of the Cu^{2+} detection when using this sensor is based on the value of the solubility product constant (K_{sp}). The values of the solubility product constant of the ZnS and CuS are 2.93×10^{-25} and 8.0×10^{-37} [15]. The relatively large difference of the solubility product constant between the ZnS and CuS constitutes the driving force for the ion exchange. Hence, when inserting the paper ZnO@ZnS core–shell NPs–based sensor in aqueous solution containing Cu ions, the Cu ions will remove the Zn in the ZnS and form CuS instead [15]. Investigating the performance of this paper sensors based on ZnO@ZnS core–shell NPs have indicated that the paper sensor is highly sensitive and can operate in a relatively large pH value. Further selectivity experiments have indicated that the present paper sensor is highly sensitive to Cu ions as intended. The lower limit of detection was found to be 15 μM (~0.96 ppm). This value is less than the international allowed level of copper ions in drinking water.

As shown by the color change in Fig. 6.7, the ZnO@ZnS core–shell NPs based sensor is highly selective for the detection of the cupper ions [15].

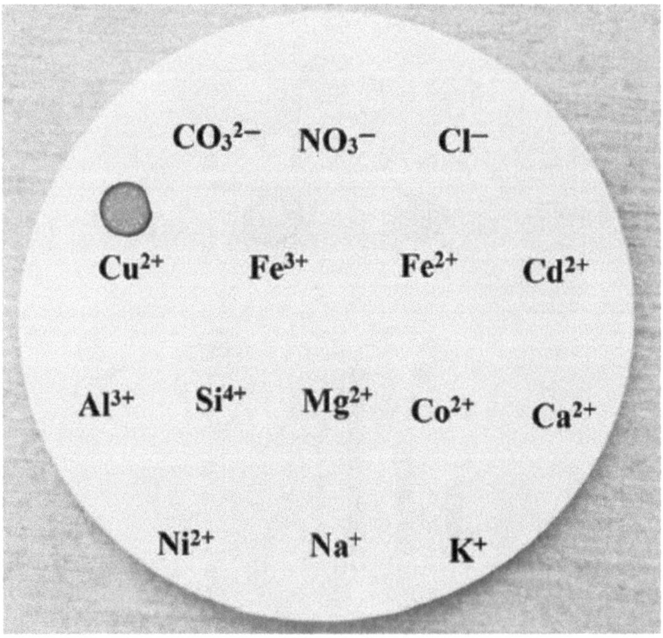

Figure 6.7 Selectivity of the ZnO@ZnS core–shell NPs–based sensor is clearly shown above as a color change when exposed to different metallic ions [15]. *NPs*, Nanoparticles.

The presented colorimetric self-powered paper sensor indicates again the potential of the low-temperature chemical synthesis of nanostructures and the possibility to develop simple, reliable, and low-cost highly sensitive and selective sensors.

6.3 New sustainable energy related applications

The population of the earth is growing fast and by the year 2050 it is expected to double. This means that the energy needs will also double. With the depletion of fossil fuel, which is the main energy source at the moment, there is an urgent need to develop a renewable energy source that is environmentally friendly due to the negative impact due to climate change and the source(s) to be developed should also be sustainable. There is great hope that nanotechnology can provide a solution to lead to the emergence of such renewable and sustainable energy source. The targeted energy source using nanomaterials is expected to be efficient and possess large electrical capacity production and/or storage. In addition, it is desired to be a pollution-free source. Energy harvesting using nanomaterials is one of the fields of interest that has gained global interest in many laboratories. The two largest sustainable energy sources are the sun and the other is the ambient mechanical energy. In this section and in the next section different examples of innovative nanomaterials and devices will be demonstrated.

Here we present two examples of utilizing the ambient mechanical energy by using the low-temperature chemical approach. In the first example, ZnO nanorods were grown at 90°C on a normal conventional text paper, as will be described below. It is known that piezoelectric materials can convert mechanical energy to electrical energy. ZnO, due to its crystal structure and arrangement of atoms in the lattice, possesses a relatively large piezoelectric effect. Using ZnO nanorods, nanopiezoelectric generators were first demonstrated by Z.L. Wang and his research group at Georgia Tech., United States. [16]. On a conventional high-flexibility paper (Invercote G from Holmen, Sweden) substrate, ZnO nanorods were synthesized by first depositing a metallic layer of chrome/silver (10/50 nm) followed by the deposition of a ZnO NP seed layer [17]. The synthesis was achieved in two steps using the standard ZnO well aligned nanorods synthesis approach described earlier [18,19]. After the growth, the growth solution, which was a milky in nature, was utilized to extract the residue which is composed of freestanding ZnO nanorods by centrifugation. The extracted ZnO nanorods powder was then left to dry and was used later. To fabricate the handwriting-driven ZnO-based nanogenerator, 0.1 g of the filtered ZnO powder was added to 1 mL of PVDF-TrFE to form an ink. By adding the ZnO nanopowder to the PVDF-TrFE the stability of the piezoelectric nanogenerator can be stabilized [20,21]. Using a stencil drop method, a small drop of this ink was applied to the surface of the ZnO nanorods grown on the piece of paper described above. Then the paper was cut into two pieces and the two pieces were attached to each other face to face.

Then the two edges of the paper were connected to an electrical measurement instrument using a crocodile clamp. Fig. 6.8 shows different photographs of the nanogenerator fabricated as described above. In Fig. 6.8A a photograph of the final device is shown, while in Fig. 6.8B a schematic diagram of the layers of the devices is shown. In Fig. 6.8C and D photographs of the piezoelectric handwriting-driven nanogenerator is shown during measurements. Two writing speed modes were investigated, these were high (fast) and low (slow) writing modes.

As can be seen in Fig. 6.9 the pressure from handwriting has generated a relatively high voltage output and the values reached a maximum of 4.8 and 2.0 V for fast and slow handwriting, respectively [17]. The maximum power was estimated to be about 1.3 mW/mm^2. Such piezoelectric paper

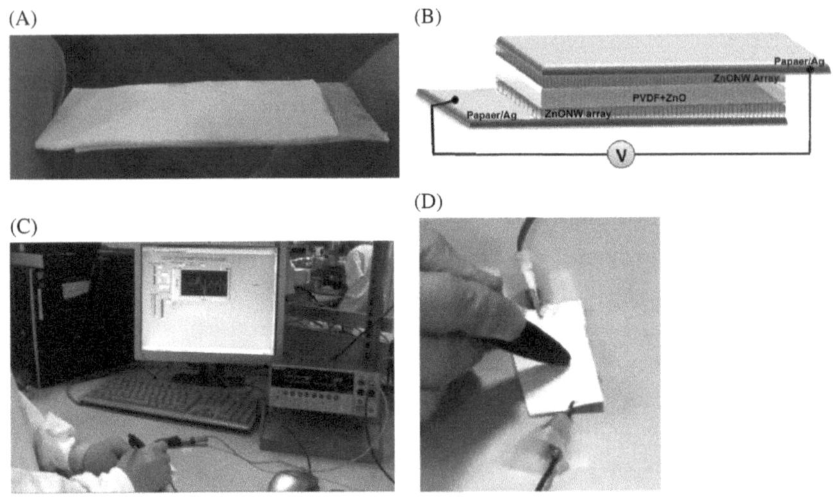

Figure 6.8 (A) A photograph showing the final piezoelectric nanogenerator, (B) a schematic diagram showing the device layers, (C) and (D) photographs showing the device while measurement [17].

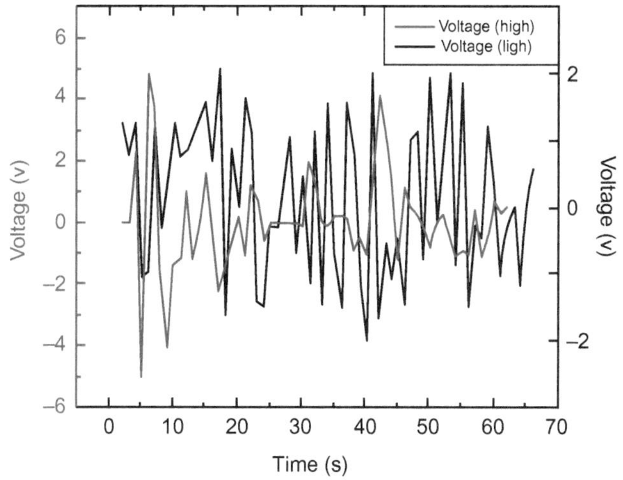

Figure 6.9 The output voltage achieved from the handwriting-driven piezoelectric nanogenerator fabricated on a paper substrate as a function of time at slow and fast speed handwriting [17].

based on ZnO nanorods grown by the low-temperature chemical approach can be used to develop smart electronic programmable paper for different applications, for example, signature verification.

The second example to be demonstrated is the use of the piezoelectric property of ZnO to develop a self-powered anisotropic direction sensor [22]. The synthesis of ZnO nanorods proceeded using the standard hydrothermal method described in the previous section using the two-step process, with the first step being the covering of the substrate front and back surfaces with ZnO relatively small NPs, followed by growth of the nanor ods [22]. Here a plastic substrate covered with PEDOT:PSS on both surfaces was used, and the substrate was inserted vertically inside the synthesis beaker.

When stacking different piezoelectric ZnO nanogenerators in series or in parallel configurations the performance can be enhanced [23,24]. Hence as can be seen in Fig. 6.10, we grew ZnO nanorods on both sides of the PEDOT:PSS-covered plastic.

Figs. 6.10 and 6.11 show a schematic diagram and a photograph of the two-sided anisotropic direction sensor, and Fig. 6.11 shows different possible bending positions and the associated harvested piezoelectric potential. As can be seen in Fig. 6.11, by bending the sensor in different directions, opposite polarity voltage is developed on the two sides of the sensor. In this way, the configuration of the two-sided ZnO nanorods configuration

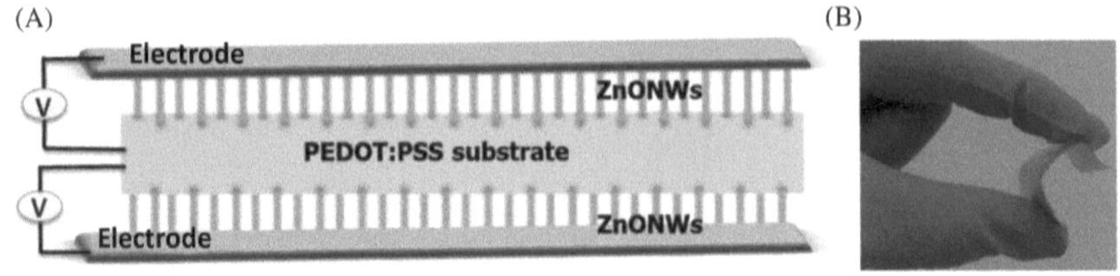

Figure 6.10 (A) Schematic diagram showing the double-sided ZnO anisotropic direction sensor and (B) a photograph of the same sensor grown on flexible plastic substrate [22].

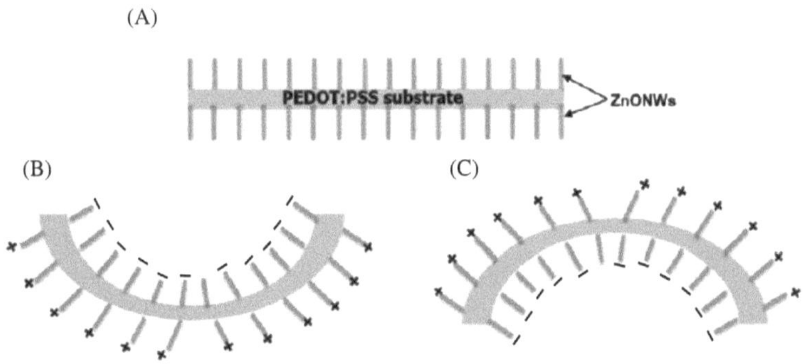

Figure 6.11 (A)–(C) Schematic diagram showing different bending situations and the associated harvested piezoelectric potential [22].

can be utilized as a direction sensor. Electrical measurements were performed by bending the sensor in different directions and measuring the harvested potential polarity on both sides.

In this example, the ZnO nanorods were grown vertically on both sides of the PEDOT:PSS-covered plastic substrate, then upon bending the top ZnO nanorods acquire positive voltage when it is compressively strained and negative voltage polarity when it is subjected to tensile strain [22]. The output voltage acquired upon bending this two-sided direction sensor is shown in Fig. 6.12. The output power delivered by bending the two-sided ZnO nanorods direction sensor fabricated on PEDOT:PSS-covered flexible plastic substrate was found to be 4.44 mW/cm^2 [22]. This example shows that using the low-temperature chemical synthesis of nanostructures a sensitive and robust direction sensor has been developed. Such a sensor can find a place as an efficient component in many applications, for example, cars and security systems.

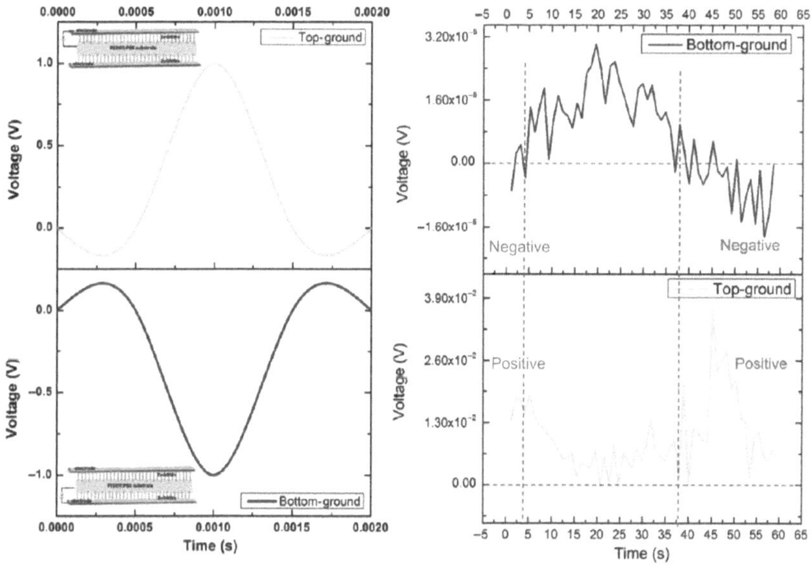

Figure 6.12 (Left) The acquired output voltage measured for a period of milliseconds from the top (green) and bottom (blue) sides (the insert shows the corresponding measured side in each case). (Right) the generated output voltage form both sides when compressing and releasing the plastic sensor for a period of 60 s [22].

6.4 New emerging photocatalysis applications

As is well-known, the consumption of energy worldwide is expected to double by the year 2050. With the depleting fossil fuel-based energy sources, there is an urgent need to develop innovative alternatives that rely on other sustainable energy sources. The sun, which is the main source of life on Earth and the direct or indirect source of all of our energy renewable systems, is one of the alternatives to be utilized by using its sustainable energy. This is due to the relatively huge energy coming to us on the Earth from the sun. Today the sun's energy is used in many applications like photovoltaic cells and thermal convertors. It is well-known that the sun's radiation is composed of visible light, infrared radiation, and ultraviolet radiation. The visible light, which is radiation with a wavelength between 0.4 and 0.75 μm, constitutes 46% of the sun radiation, while the infrared radiation takes a share of about 47%, and the ultraviolet light occupies the remaining 7%. Ultraviolet radiation with the shortest wavelength and highest accompanying energy has been used continuously in the industry to complete many photocatalytic processes. The ultraviolet radiation used in our industry for photocatalysis today is obtained artificially using relatively high-energy consumable methods. In this connection the nanomaterials with their high surface area to volume ratio have been proven to achieve almost the same photocatalytic activity of other materials utilizing the ultraviolet radiation but with the difference of being activated by the visible radiation. This means that nanomaterials can be highly catalytic using a visible radiation of wavelengths between 0.4 μm and 0.75 μm. This fact has opened a research area which has attracted global interest in many laboratories worldwide because of the possibility of utilizing the visible sustainable radiation of the sun to perform many photocatalytic processes instead of using ultraviolet radiation.

Another energy related problem today is the accompanying carbon emissions when using the conventional energy sources that we utilize at the moment. As is well-known, carbon emissions are connected to global warming and the deterioration of the climate on Earth. In this connection, it is well-known that hydrogen as an energy source can be used to deliver electricity, power cars, and give heating to our homes in the winter. However, the conventional methods available today to produce hydrogen rely on fossil-based fuel, in addition to being relatively expensive. This creates a dilemma. At present the hydrogen produced with the available conventional methods will not solve the carbon emissions obstacle. Here again nanomaterials are providing hope for developing electrodes that can split water and produce hydrogen using sustainable visible light from the sun instead of from fossil fuels. It is well-known that hydrogen is a gas, and hence there is a challenging question concerning the storage and security. In addition, most of our systems today are based on liquid fuel tanks, but still to produce hydrogen can be quite strategically important if we can produce it from a clean sustainable source, like the sun's visible light. This hydrogen could be used easily to produce methanol in an efficient economic and environment-friendly manner. In this connection we will present here two examples based on the use of the low-cost low-temperature chemical synthesis for developing photocatalytic efficient materials and for splitting water to produce hydrogen.

Using the low-temperature chemical synthesis, a disruptive plasmonic ZnO/graphene/Ag/AgI nanocomposite was produced for solar-driven photocatalysis [25]. To achieve such a nanocomposite, the hydrothermal low-temperature chemical synthesis was combined with ultrasonication. For the synthesis of this nanocomposite, analytical grade chemicals were purchased from Sigma Aldrich and were used without further purification. The chemicals and compounds used were multilayer graphene (GR) powder, zinc acetate dihydrate, potassium hydroxide, silver nitrate, and sodium iodine. For the preparation of all aqueous solutions, deionized water was used [25]. Fig. 6.13 shows a schematic diagram displaying the steps performed to obtain the ZnO/GR/Ag/AgI nanocomposite.

The Zn/GR/Ag/AgI plasmonic nanocomposite NP developed here was used as a photocatalyst to degrade Congo red (CR) dye. In addition, degradation experiments were also performed for ZnO, ZnO/GR, and ZnO/Ag/AgI as reference samples to compare them with the ZnO/GR/Ag/AgI nanocomposite. The photodegradation experiments were conducted using a solar simulator giving a radiation equivalent to 1 sun. The results indicated that the degradation efficiency of the Zn/GR/Ag/AgI was superior to the other reference samples and reached 90% after 60 minutes of degradation. This is explained by the synergetic effect between the ZnO/GR and the Ag/AgI [25]. The calculated degradation efficiencies of the ZnO, ZnO/GR, ZnO/Ag/AgI, and ZnO/GR/Ag/AgI are

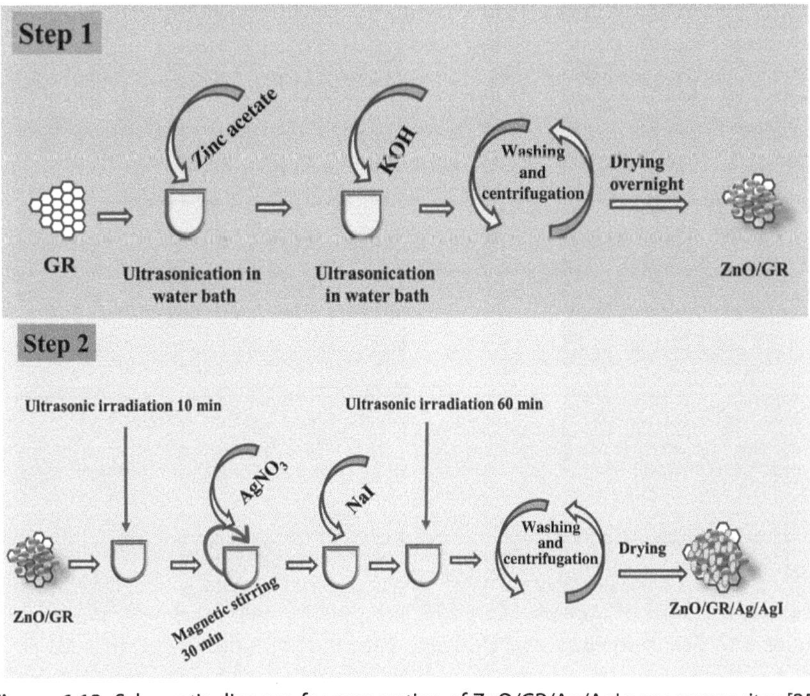

Figure 6.13 Schematic diagram for preparation of ZnO/GR/Ag/AgI nanocomposites [25].

shown in Fig. 6.14A. As is clearly shown, in the absence of any photocatalyst the photodegradation of the CR is negligible implying that the CR is stable during extended periods of irradiation. However, when ZnO NPs, or ZnO/GR nanocomposite were used an enhancement of the photodegradation is clearly observed as shown in Fig. 6.14A. Specifically, the addition of the GR has led to an increase in the degradation efficiency to about 75% compared to 58% for pure ZnO. This is attributed to the inhibition of the recombination of the visible light–photogenerated e^-/h^+ pairs [25]. Also as can be clearly seen in Fig. 6.15A, the photocatalytic efficiency of the ZnO/Gr/Ag/AgI depends strongly on the Ag/AgI weight ratio. For a 20% weight ratio the efficiency reached 90% [25]. However, when the weight ratio is increased to 30% the efficiency drops down to 60%. This is explained by the fact that excess Ag/AgI creates recombination centers which reduce the active e^-/h^+ pairs available for the photocatalysis.

The mechanism involved in the visible light-driven photodegradation of the CR using the ZnO/GR/Ag/AgI (20%) was investigated by using

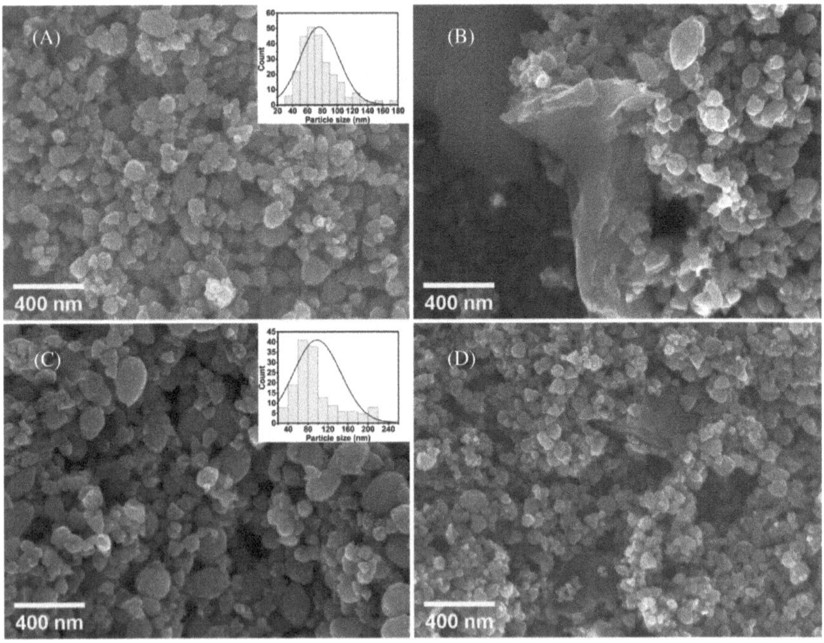

Figure 6.14 Scanning electron microscope images of (A) pristine ZnO NPs, (B) ZnO/GR, (C) ZnO/Ag/AgI, and (D) ZnO/GR/Ag/AgI (20%) nanocomposites. The insets show the size distribution of the nanoparticles [25].

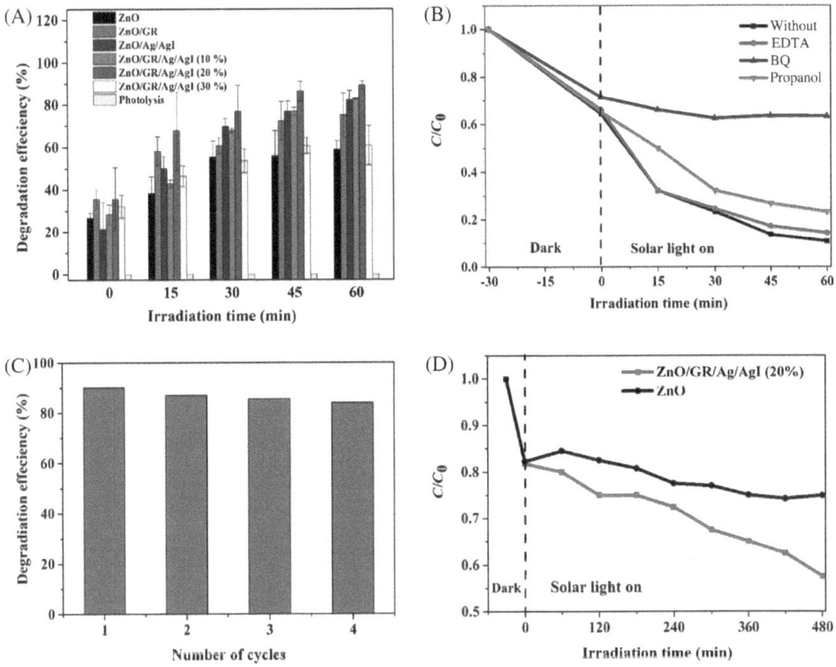

Figure 6.15 (A) Photodegradation efficiency of CR dyes using ZnO, ZnO/GR, ZnO/Ag/AgI and ZnO/GR/Ag/AgI photocatalysts with different weight percentage of Ag/AgI, (B) photocatalytic activities of CR with ZnO/GR/Ag/AgI (20%) nanocomposite in the presence of different scavengers, (C) recycling of ZnO/GR/Ag/AgI (20%) nanocomposite during degradation of CR, and (D) photodegradation of phenol with ZnO and ZnO/GR/Ag/AgI (20%) nanocomposite [25]. *CR*, Congo red.

trapping experiments. These were performed in order to identify the reactive species that led to the photodegradation mechanism. For this purpose, ethylenediaminetetraacetic acid (EDTA), benzoquinone (BQ), and 2-propanol were used as scavengers for h +, superoxide radicals, and hydroxyl radicals, respectively [26–29]. Fig. 6.15B shows the results of adding the compounds for trapping experiments. As can be seen, the addition of the BQ led to a suppression of the photodegradation process due to the quenching of the superoxide radicals, which is an indication that the superoxide is the main radical that causes the photodegradation under visible radiation in these experiments. In addition, the addition of the 2-propoanol indicated a partial inhibition which implies that the hydroxyl group has an important role in the degradation of the CR using visible light. As a surprise the addition of the EDTA did not affect the degradation at the beginning of the experiments. Nevertheless, after

30 minutes, quenching of the holes was observed to increase [25]. The high efficiency and excellent visible light–driven photodegradation observed from the ZnO/GR/Ag/AGI nanocomposite necessitated and encouraged a stability study as such a nanocomposite showed good potential. After the photodegradation experiment, the nanocomposite was again collected and used after cleaning for other tests. The same nanocomposite was used for four successive photodegradation experiments. The result is shown in Fig. 6.15C. As can be seen, the reduction of the efficiency is minor after four successive uses under visible light. This indicates an excellent stability of the ZnO/GR/Ag/AgI as a visible light–driven photocatalyst [25]. The high efficiency of the ZnO/GR/Ag/AgI under visible sunlight has encouraged the performing of further experiments to further study the photosensitization of the CR under visible light. For this the photocatalytic performance of the present nanocomposite was investigated for the degradation of phenol which is known to be insensitive to visible light. A 100 mL solution of phenol (0.02 g/L) was added to 0.05 g of the nanocomposite and was stirred for 30 minutes in the dark. For comparison also ZnO NPs were prepared and added to the phenol for comparison. Then the photocatalytic degradation was investigated and the results (shown in Fig. 6.15D) indicated that the ZnO/GR/Ag/AgI is superior in its visible light–driven photodegradation. This implies that the effect of sensitization on the degradation efficiency is negligible for both phenol and CR [25]. This example shows the potential that can be provided by the low-temperature chemical synthesis of nanostructures to contribute to the development of efficient processes based on a renewable energy source like the sun's visible radiation.

The second example to be highlighted here is concerning the development of a renewable clean energy source, which is hydrogen again using the sun as a renewable energy source by utilizing nanoelectrodes developed using the low-temperature chemical approach. Hydrogen was discovered and considered as an element as early as 1716 by Henry Cavendish. Hydrogen provides the highest energy content of any common fuel considering its weight; however, it provides the lowest energy considering its volume. Hydrogen is today used as a fuel in few areas, since the present fabrication processes to produce hydrogen are relatively expensive. Due to this and the fact that hydrogen is a gas and hence light in weight, it is mainly used in missiles and for fueling space craft missions to reduce the weight of the fuel. The French novelist Jules Verne who wrote a book entitled the "The Mysterious Island," which was published in 1874 and was the basis for a film, wrote *I believe that water will one day be employed as fuel, that hydrogen and oxygen which constitute it, used singly or together, will furnish an inexhaustible source of heat and light, of an intensity of which coal is not capable.* Although this was written more than 100 years ago and Jules Verne was famous as a science fiction writer, today this has become a reality! Splitting water into hydrogen and oxygen using different methods is in fact a hot research topic and many laboratories worldwide are concentrating on developing an efficient, low-cost, and durable process to provide hydrogen to be used as a clean energy source. One of the methods and efforts to provide hydrogen in an efficient way is the use of electrochemical photocatalysis using nanoelectrodes synthesized by the low-temperature chemical approach [30]. Electrolysis is the process by which water is decomposed into oxygen and hydrogen. To provide an efficient and acceptable process this electrolysis should be derived by a renewable energy source, then it can be considered as a breakthrough for providing clean energy source. Although this is a challenging task there are many efforts to reach the target. Among them is the utilization of nanostructures synthesized by the low-temperature chemical approach. Following this approach, the researchers aim to develop an efficient and durable electrode to act as a photoelectrochemical cell. To have a renewable energy source to provide such photoelectrochemical, the use of visible sunlight as the driving source will be highlighted.

Nowadays the state-of-the-art electrodes for oxygen evolution reaction (OER) and hydrogen evolution reaction (HER) through water splitting are based on Pt and IrO_2/RuO_2 [30,31]. However, these metals are relatively expensive and scaling up the hydrogen production would be a barrier for efficient low-cost production. A recent successful attempt to develop an electrode based on a process which can be scaled up that is based on a low-cost approach has been recently published [32]. Nickel–iron layer double hydroxide (NiFeLDH) was deposited on cobalt oxide (Co_2O_3) nanowires using a chemical low-temperature approach. In this work the NiFeLDH covered the nanowires and formed a core–shell configuration [32]. Fig. 6.16 below shows scanning electron microscope images of the NiFeLDH/Co_2O_3 core–shell nanowires. The details of the synthesis process can be found in Ref. [32].

Due to the modification of the surface of the Co_2O_3 by the NiFeLDH the number of active sites has increased. In turn this has

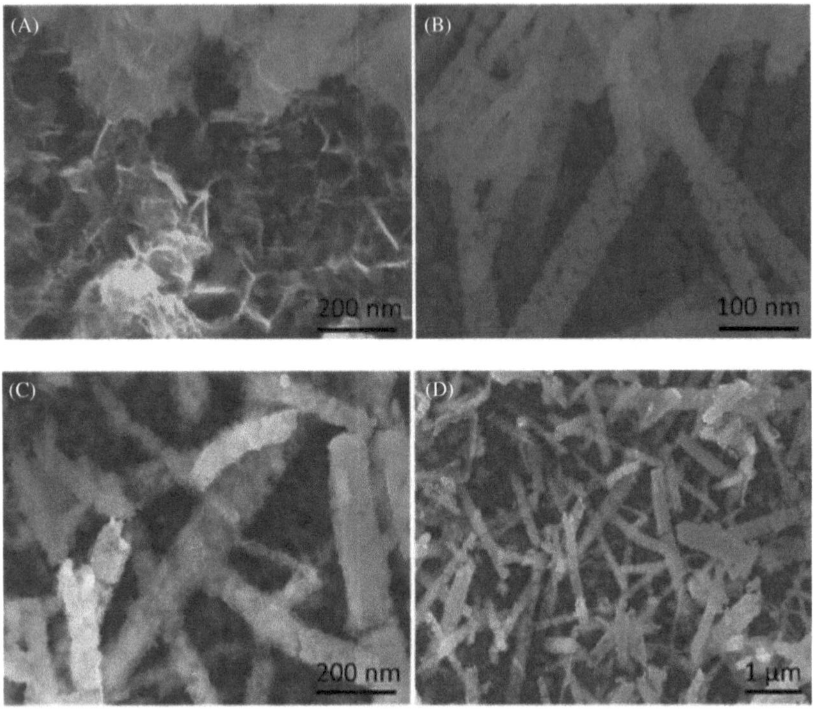

Figure 6.16 Scanning electron microscope images showing (A) pure NiFeLDH, (B) pure Co_3O_4, and (C) and (D) NiFeLDH functionalized Co_3O_4 nanowires [32].

increased the efficiency of the Co_2O_3 nanowires for water splitting. Investigating these NiFeLDH/Co_2O_3 core–shell nanowires, it was found that the deposition of the LDH has led effectively to the appearance of a synergetic effect and it has a double bifunctionality for water splitting in 1 M KOH aqueous solution. At a potential of −0.303 and by using a reversible hydrogen electrode a current density of 10 mA/cm^2 was observed. While for the OER a current density as high as 40 mA/cm^2 was achieved at a voltage of 1.49 V [32]. The stability of this core–shell electrode was found to be 15 hours for both the HER and OER. Using electrochemical impedance spectroscopy this electrode has shown the fact that by depositing the NiFeDLH onto the surface of the Co_2O_3 the charge transfer increases threefold for both the HER and the OER [32]. These observations indicate that the suggested electrode is having an efficient water-splitting performance.

6.5 Low-temperature nanofabrication: challenges and future prospects

As has been presented in the different chapters of this book, low-temperature nanofabrication based on chemical routes is providing a variety of high-quality nanostructures. The research efforts during the last decade using the bottom-up chemical low-temperature methods to fabricate multifunctional devices have laid the groundwork for a wealth of extra knowledge due to the fascinating nature and characteristics of nanostructures. The advancements of nanofabrication techniques are hoped to provide technological breakthroughs for many applications, as conventional microelectronics have reached their maximum benefits to humanity. Hence, the era of nanoelectronics is the hope for the expected breakthrough. For example, for the area of data storage, researchers hope that one day a single data bit can be stored in a single atom! The expected breakthrough includes areas such as nanomedicine for diagnostics and drug delivery, which could benefit from nanoscience knowledge.

Although the gains of the last decade in the area of nanoscience and technology have been rich, still many challenges are to be overcome before we can have nanofabrication as a standard tool for fascinating new products. In general, the bottom-up techniques discussed here have shown that it is possible to obtain complex nanostructures that are difficult to fabricate using the conventional top-down methods. As has been discussed earlier in this book, top-down techniques for nanofabrication are still series-based products and mass production will be difficult to achieve at a reasonable cost. However, in general, the present trend in nanoscience is not only that new prospects are provided to the industry, but that hugely useful knowledge is being gained every day. Examples of this new useful knowledge are clear in the fields of medicine and in solving energy needs. The knowledge known to researchers today using the low-temperature chemical synthesis, although they have provided excellent multifunctional nanostructures, still needs optimization for large-scale production. Such a task will still need some research efforts and then such methods which are low-cost can be utilized for industrial products. It is evident that to provide scaled-up products that will constitute a real breakthrough, both top-down conventional techniques should be integrated with the state-of-the-art bottom-up techniques. In this way, conventional lithography can, for example, be integrated with bottom-up approaches to provide, for example, very large-scale integrated nanoelectronics circuits. With the present trend, such steps can be a reality in the near future if funds are allocated for it.

References

[1] Kamran ul Hasan, O. Nur, M. Willander, Appl. Phys. Lett. 100 (2012) 211104.
[2] A. Wadeasa, O. Nur, M. Willander, Nanotechnology 20 (2008) 065710.
[3] S. Nour Eiman, C.O. Chey, M. Willander, O. Nur, Nanotechnology 26 (2015) 095502.
[4] A. Khan, M.A. Abbasi, J. Wissting, O. Nur, M. Willander, Phys. Status Solidi RRL 7 (2013) 980.
[5] H.J. Koo, J. Seo, T. Lee, Thin Solid Films 524 (2012) 1.
[6] M.H. Asif, A. Fulati, O. Nur, M. Willander, C. Brännmark, P. Strålfors, et al., Appl. Phys. Lett. 95 (2009) 023703.
[7] M.H. Asif, S.M. Ali, O. Nur, M. Willander, C. Brännmark, P. Strålfors, et al., Biosens. Bioelectron. 15 (2010) 2205.
[8] B. Draznin, K.E. Sussman, R.H. Eckel, M. Kao, T. Yost, N.A. Sherman, J. Clin. Invest. 82 (1988) 1848.
[9] T. Takahashi, E. Neher, B. Sakmann, Proc. Natl. Acad. Sci. U.S.A. 84 (1987) 5063.
[10] A.S.N. Raphael, R.E. Özel, W.H. Mak, M. Mulato, B. Singaram, N. Pourmand, Nano Lett. 161194 (2016).
[11] A.A. Pezzulo, M.H. Asif, M. Willander, J. Anal. Bioanal. Tech. 7 (2011). Available from: https://doi.org/10.4172/2155-9872.S7-002.
[12] S.I. Börjesson, U.H. Englund, M.H. Asif, M. Willander, F. Elinder, J. Biol. Chem. 46 (2011) 39823.
[13] A. Sadollahkhani, I. Kazeminezhad, L. Jun, O. Nur, L. Hultman, M. Willander, RSC Adv. 4 (2014) 36940.
[14] A. Sadollahkhani, O. Nur, M. Willander, I. Kazeminezhad, V. Khranovskyy, M.O. Eriksson, et al., Mater. Chem. Phys. 163 (2015) 485.
[15] A. Sadollahkhani, A. Hatamie, O. Nur, M. Willander, B. Zargar, I. Kazeminezhad, Appl. Mater. Interfaces 6 (2014) 17694.
[16] Z.L. Wang, J.H. Song, Science 312 (2006) 242.
[17] E.S. Nour, M.O. Sandberg, M. Willander, O. Nur, Nano Energy 9 (2014) 221.
[18] C. Pacholski, A. Kornowski, H. Weller, Angew. Chem. Int. Ed. 41 (2002) 1188.
[19] L. Vayssieres, Adv. Mater. 15 (2003) 464.
[20] Y. Hu, Y. Zhang, C. Xu, G. Zhu, Z.L. Wang, Nano Lett. 10 (2010) 5025.
[21] M. Lee, C.-Y. Chen, S. Wang, S.N. Cha, Y.J. Park, J.M. Kim, et al., Adv. Mater. 24 (2012) 1759.
[22] E.S. Nour, C.O. Chey, M. Willander, O. Nur, Nanotechnology 26 (2015) 095502.
[23] R.S. Yang, Y. Qin, C. li, L.M. Dai, Z.L. Wang, Appl. Phys. Lett. 94 (2009) 022905.
[24] R.S. Yang, Y. Qin, L.M. Dai, Z.L. Wang, Nat. Nanotechnol. 4 (2009) 34.
[25] A. Adam Rania, E. Chalangar, M. Pirhashemi, G. Pozina, X. Liu, J. Palisaitis, et al., RSC Adv. 9 (2019) 30585.
[26] T.T.H. Tran, H. Kosslick, A. Schulz, Q.L. Nguyen, Adv. Nat. Sci. Nanosci. Nanotechnol. 8 (2017) 015011.
[27] L. Hou, H. Zhang, X. Xue, Sep. Purif. Technol. 84 (2012) 147.
[28] A.L. Teel, R.J. Watts, J. Hazard. Mater. 94 (17) (2002) 9.
[29] D.A. Reddy, J. Choi, S. Lee, R. Ma, K.T. Kyu, RSC Adv. 5 (2015) 67394.
[30] N. Cheng, S. Stambula, D. Wang, M.N. MBanis, J. Liu, A. Riese, et al., Nat. Commun. 7 (2016) 13638.
[31] G. Li, S. Li, J. Ge, C. Liu, W. Xing, J. Mater. Chem. A 5 (2017) 17221.
[32] A. Aneela Tahira, Z.H. Ibupoto, M. Vagin, U. Aftab, M.I. Abro, M. Willander, et al., Catal. Sci. Technol. 9 (2019) 2879.

Index

Note: Page numbers followed by "*f*" and "*t*" refer to figures and tables, respectively.

T

FSC
www.fsc.org
MIX
Papier | Fördert gute Waldnutzung
FSC® C083411

Kolibri 360 GmbH
Ferdinand-Jühlke-Straße 7
99095 Erfurt, Deutschland
produktsicherheit@kolibri360.de